# Number theory and discrete geometry

## Ramanujan Mathematical Society Lectures Notes Series

Vol. 1: Number theory

Vol. 2: The Riemann zeta function and related themes

Vol. 3: Formal language aspects of natural computing

Vol. 4: Commutative algebra and combinatorics

Vol. 5: Convexity in discrete structures

Vol. 6: Number theory and discrete geometry

Vol. 7: Discrete mathematics

Vol. 8: Lectures on operator theory

Vol. 9: Essays on geometric group theory

Vol. 10: Techmüller theory and moduli problems

*Ramanujan Mathematical Society*

Lecture Notes Series

Volume 6

# Number theory and discrete geometry

Proceedings of the international conference
on number theory and discrete geometry,
Panjab University, Chandigarh,
December 2005

*Volume editors*

R. Balasubramanian
S. G. Dani
P. M. Gruber
R. J. Hans-Gill

International Press
www.intlpress.com

Ramanujan Mathematical Society
Lecture Notes Series, Volume 6
Number theory and discrete geometry

Volume Editors:
R. Balasubramanian
S. G. Dani
P. M. Gruber
R. J. Hans-Gill



Published in 2010 by International Press, Somerville, Massachusetts, U.S.A. under license from the Ramanujan Mathematical Society.

This work is also published in the Republic of India, exclusively by the Ramanujan Mathematical Society.



ISBN 978-1-57146-191-9

Typeset using the LaTeX system.
Printed in the United States of America.

# Foreword

When I reflect on the great number theorists of India, the names that come to my mind are S. Ramanujan, S. S. Pillai, S. D. Chowla and R. P. Bambah. Chowla and Bambah have also contributed greatly by building good schools around them. In the context of the general drift towards many newer topics, building up a school in a hard area like classical number theory (which has been laboured on for ages) is indeed a difficult task. It is to the immense credit of Prof. R. P. Bambah to have, in addition to prolific and deep contributions of his own, in the areas of Geometry of Numbers, Packings and Coverings, Discrete Geometry, Diophantine Approximations and classical Number Theory, built such a school of high standard.

Professor Bambah has played a pivotal role in the development of the theory of coverings, and his work has triggered a good deal of further research on the theme. There are many significant results in the area proved by him and by others benefiting from his guidance. Proofs of Watson's conjectures and other related conjectures on non-homogeneous quadratic forms, progress in finite coverings, in the results on covering densities for spheres and the well-known conjecture of Minkowski are some notable examples.

Professor Bambah holds the distinction of having close association with many celebrated mathematicians like H. Davenport, L. J. Mordell, P. Erdös, K. F. Roth, H. Zassenhaus, and so on. He has been very interactive and has had numerous research collaborations, involving a wide spread of mathematicians: S. Chowla, H. Davenport, C. A. Rogers, A. C. Woods, H. Gupta, K. Rogers, K. F. Roth, H. Zassenhaus, V. C. Dumir, R. J. Hans-Gill, N. J. A. Sloane, I. S. Luthar, M. L. Madan, D. D. Joshi, D. B. Lahiri.

Professor Bambah is very erudite on the development of number theory. He is an authority on the history of the Manchester School built by L. J. Mordell and the Cambridge School built by G. H. Hardy and J. E. Littlewood. He is also very knowledgeable about S. Ramanujan, E. C. Titchmarsh, A. E. Ingham and many others.

I would like to recall here a simple looking result proved by Bambah and Chowla which deserves keen attention of the coming generations of mathematicians. Here it is: let $q_1, q_2, \ldots$ be the ascending sequence of natural numbers which are sums of two squares of integers. Then as $n$ tends to infinity, we have $q_{n+1} - q_n = O((q_n)^a)$, where $a = \frac{1}{4}$. Bambah and Chowla gave an $O$-constant. It is a great challenge to improve even their $O$-constant (let alone proving $o((q_n)^a)$, with $a = \frac{1}{4}$). It is a still greater challenge to prove the result with $a$ arbitrarily small.

I could go on, writing more and more about Professor Bambah. However I will conclude here, wishing him and his folk (including his family and his students) a long and happy life.

**K. Ramachandra**

# Preface

An International Conference in Number Theory and Discrete Geometry was held at the Department of Mathematics, Panjab University, Chandigarh from November 30 to December 3, 2005, in honour of Professor R. P. Bambah on the occasion of his 80th birthday. This volume represents the proceedings of the conference. Some invited speakers who could not attend the conference have also contributed to the volume.

We thank the authors for their contributions to this volume, and the referees of all papers for helping to improve the quality of the volume. We also take this opportunity to thank all those whose efforts contributed to the success of the conference, including the Organising Committee of the conference, the participants at the conference, and the faculty and administrative staff of the Mathematics Department, Panjab University who enthusiastically carried out various responsibilities for the conference. Thanks are also due to National Board for Higher Mathematics, Council for Scientific and Industrial Research, Indian National Science Academy and Panjab University, Chandigarh for financial support to the conference. We are also thankful to the Ramanujan Mathematical Society for agreeing to publish the proceedings in the RMS Lecture Notes Series sponsored by the Department of Science and Technology.

**R. Balasubramanian**
**S. G. Dani**
**P. M. Gruber**
**R. J. Hans-Gill**

# Organising Committee

| | |
|---|---|
| **Convener** | *Prof. R. J. Hans-Gill* |
| **Co-conveners** | *Prof. Madhu Raka & Prof. V. K. Grover* |
| **Members** | *Prof. A. K. Agarwal*<br>*Prof. A. K. Bhandari*<br>*Prof. V. C. Dumir*<br>*Dr. Gurmeet K. Bakshi*<br>*Prof. Sudesh K. Khanduja*<br>*Prof. D. S. Khassa*<br>*Prof. I. B. S. Passi*<br>*Dr. D. B. Rishi* |

# Sponsored by

National Board for Higher Mathematics
Council for Scientific and Industrial Research
Indian National Science Academy
Panjab University

# List of Participants

**Prof. S. D. Adhikari**, Harish Chandra Institute, Allahabad.
**Prof. A. K. Agarwal**, Panjab University, Chandigarh.
**Prof. K. T. Arasu**, Wright State University, Dayton, USA.
**Ms. Shivani Anand**, Panjab University, Chandigarh.
**Ms. Arpana**, Panjab University, Chandigarh.
**Mr. Ashish Arora**, G.N.D.U., Amritsar.
**Dr. Renu Bajaj**, Panjab University, Chandigarh.
**Dr. Gurmeet K. Bakshi**, Panjab University, Chandigarh.
**Ms. Suman Bala**, Panjab University, Chandigarh.
**Prof. R. Balasubramanium**, Institute of Mathematical Sciences, Chennai.
**Prof. R. P. Bambah**, Panjab University, Chandigarh.
**Prof. Bindu A. Bambah**, University of Hyderabad, Hyderabad.
**Prof. A. K. Bhandari**, Panjab University, Chandigarh.
**Dr. Saurabh Bhatia**, UIET, Panjab University, Chandigarh.
**Dr. Savita Bhatnagar**, Panjab University, Chandigarh.
**Dr. Chandan Bist**, Kumaon University, Nainital.
**Dr. Vikas Bist**, Panjab University, Chandigarh.
**Prof. M. N. Bleicher**, University of Wisconsin, Madison, USA.
**Prof. Jasbir Singh Chahal**, Brigham Young University, Provo, USA.
**Prof. S. G. Dani**, Tata Institute of Fundamental Research, Mumbai.
**Prof. L. Danzer**, University of Dortmund, Germany.
**Prof. V. C. Dumir**, Panjab University, Chandigarh.
**Ms. Gagan Deep**, Panjab University, Chandigarh.
**Dr. Dev Dulari**, Panjab University, Chandigarh.
**Prof. G. Fejes Tóth**, Alfréd Rényi Inst. Mathematics, Budapest, Hungary.
**Ms. Gitika**, Panjab University, Chandigarh.
**Prof. V. K. Grover**, Panjab University, Chandigarh.
**Prof. P. M. Gruber**, Technische Universitat, Austria.
**Prof. R. N. Gupta**, Panjab University, Chandigarh.
**Prof. R. J. Hans-Gill**, Panjab University, Chandigarh.
**Prof. Martin Henk**, Universität Magdeburg, Magdeburg, Germany.
**Ms. Jatinder Kaur**, Panjab University, Chandigarh.
**Mr. Jitendra Singh**, Panjab University, Chandigarh.

**Prof. S. Kanemitsu**, Kinki University, Fukuoka, Japan.
**Prof. S. K. Khanduja**, Panjab University, Chandigarh.
**Prof. D. S. Khassa**, Panjab University, Chandigarh.
**Dr. Dinesh K. Khurana**, Panjab University, Chandigarh.
**Ms. Aarti Khurana**, Panjab University, Chandigarh.
**Prof. S. K. Katre**, University of Pune, Pune.
**Ms. Sukhvinder Kaur**, Panjab University, Chandigarh.
**Ms. Harsimran**, SUSCET, Tangori.
**Dr. Shanta Laishram**, TIFR, Mumbai.
**Prof. I. S. Luthar**, Panjab University, Chandigarh.
**Prof. Sunder Lal**, Panjab University, Chandigarh.
**Ms. Rachna**, SUSCET, Tangori.
**Ms. Meenakshi**, Chitkara Inst. Engg. & Tech, Jansla, Rajpura.
**Prof. M. Ram Murty**, Queen's University, Kingston, Canada.
**Prof. V. Kumar Murty**, University of Toronto, Toronto, Canada.
**Prof. V. C. Nanda**, Panjab University, Chandigarh.
**Dr. J. C. Parnami**, Panjab University, Chandigarh.
**Prof. I. B. S. Passi**, Panjab University, Chandigarh.
**Dr. Gyan Prakash**, Harish Chandra Research Institute, Allahabad.
**Prof. Dipender Prasad**, TIFR, Mumbai.
**Prof. K. C. Prasad**, Ranchi University, Ranchi.
**Dr. Anju Pusri**, Panjab University, Chandigarh.
**Dr. Ram Krishna Pandey**, IIT, Delhi.
**Prof. K. Ramachandra**, NIAS, I.I.Sc. Banglore.
**Prof. Madhu Raka**, Panjab University, Chandigarh.
**Dr. D. B. Rishi**, Panjab University, Chandigarh.
**Mr. Sarad Sanasam**, Panjab University, Chandigarh.
**Prof. N. Saradha**, TIFR, Mumbai.
**Dr. Asha Sarin**, MCMDAV College, Chandigarh.
**Prof. K. S. Sarkaria**, Panjab University, Chandigarh.
**Mr. Anuj Sharma**, Panjab University, Chandigarh.
**Dr. Poonam Sehgal**, Panjab University, Chandigarh.
**Dr. Ranjeet Sehmi**, Panjab Engineering College, Chandigarh.
**Dr. Dibagh Singh**, Panjab University, Chandigarh.
**Prof. K. Srinivas**, Institute of Mathematical Sciences, Chennai.
**Dr. Sucheta**, Panjab Engineering College, Chandigarh.
**Prof. R. Sujatha**, TIFR, Mumbai.
**Dr. Kalpna Sharma**, UIET, P.U., Chandigarh.

**Dr. Suryaramana**, Harish Chandra Research Institute, Allahabad.

**Dr. Amitabha Tripathi**, IIT, New Delhi.

**Prof. S. K. Tomar**, Panjab University, Chandigarh.

**Prof. H. K. L. Vasudeva**, Panjab University, Chandigarh.

**Dr. Vanita Verma**, Panjab University, Chandigarh.

**Dr. Keerti Vardhan**, DES, Panjab University, Chandigarh.

**Prof. Jörg M. Wills**, Universität Siegen, Siegen, Germany.

# Publications of Professor R. P. Bambah


1. Number theory (Edited by R. P. Bambah, V. C. Dumir and R. J. Hans-Gill) Trends in Mathematics, *Birkhäuser Verlag, Basel* (2000), viii+527 pp.
2. Non-homogeneous problems: conjectures of Minkowski and Watson (with Dumir, V. C.; Hans-Gill, R. J.) *Number theory*, 15–41, Trends Math., *Birkhäuser, Basel* (2000).
3. Packings and coverings, *Indian J. Pure Appl. Math.* **30** (1999), no. 10, 1063–1072.
4. Chowla, the mathematics man, *Math. Student* **67** (1998), no. 1–4, 153–161.
5. Diophantine inequalities (with Dumir, V. C.; Hans-Gill, R. J.) *Proc. Nat. Acad. Sci. India Sect. A* **68** (1998), no. 2, 101–114.
6. The conjectures of Minkowski, Watson and others *Math. Student* **65** (1996), no. 1–4, 176–178.
7. On a problem of G. Fejes Tóth (with Woods, A. C.) K. G. Ramanathan memorial issue. *Proc. Indian Acad. Sci. Math. Sci.* **104** (1994), no. 1, 137–156.
8. On an analogue of a problem of Mordell (with Dumir, V. C.; Hans-Gill, R. J.) *Studia Sci. Math. Hungar.* **21** (1986), no. 1–2, 135–142.
9. Positive values of non homogeneous indefinite quadratic forms (with Dumir, V. C.; Hans-Gill, R. J.) *Topics in classical number theory, Vol. I, II (Budapest, 1981),* 111–170, Colloq. Math. Soc. János Bolyai, 34, *North-Holland, Amsterdam* (1984).
10. Positive values of non homogeneous indefinite quadratic forms II (with Dumir, V. C.; Hans-Gill, R. J.) *J. Number Theory* **18** (1984), no. 3, 313–341.
11. On a conjecture of Jackson on non homogeneous quadratic forms (with Dumir, V. C.; Hans-Gill, R. J.). *Number Theory* **16** (1983), no. 3, 403–419.
12. On a problem of Ryskov concerning lattice coverings (with Sloane, N. J. A.) *Acta Arith.* **42** (1982/83), no. 1, 107–109.
13. Srinivasa Ramanujan Medal Lecture—1979. Number theory—many challenges, some achievements. *Proc. Indian Nat. Sci. Acad. Part A* **46** (1980), no. 2, 109–118.
14. Minkowski's conjecture for $n = 5$; a theorem of Skubenko (with Woods, A. C.) *J. Number Theory* **12** (1980), no. 1, 27–48.
15. Mathematics and society *Bull. Math. Assoc. India* **11** (1979), no. 1–2, 10–18.
16. Covering by star domains (with Dumir, V. C.; Hans-Gill, R. J.) *Indian J. Pure Appl. Math.* **8** (1977), no. 3, 344–350.
17. On the product of three inhomogeneous linear forms (with Woods, A. C.) *Number theory and algebra*, pp. 7–18. *Academic Press, New York* (1977).
18. On a theorem of Dyson. Collection of articles dedicated to K. Mahler on the occasion of his seventieth birthday (with Woods, A. C.) *J. Number Theory* **6** (1974) 422–433.
19. Geometry of numbers, packing and covering and discrete geometry, Volume dedicated to the memory of V. Ramaswami Aiyar. *Math. Student* **39** (1971), 117–129 (1972)

20. On a problem of Danzer (with Woods, A. C.) *Pacific J. Math.* **37** (1971), 295–301.
21. The thinnest double lattice covering of three-spheres (with Woods, A. C.) *Acta Arith.* **18** (1971) 321–336
22. On plane coverings with convex domains (with Woods, A. C.) *Mathematika* **18** (1971) 91–97.
23. Packing and covering. *Math. Student* **38** (1970) 133–138.
24. On minimal density of plane coverings by circles (with Woods, A. C.) *Acta Math. Acad. Sci. Hungar* **19** (1968) 337–343.
25. The covering constant for a cylinder (with Woods, A. C.) *Monatsh. Math.* **72** (1968) 107–117.
26. Convex bodies with a covering property (with Woods, A. C.) *J. London Math. Soc.* **43** (1968) 53–56.
27. On the minimal density of maximal packings of the plane by convex bodies (with Woods, A. C.) *Acta Math. Acad. Sci. Hungar.* **19** (1968) 103–116.
28. Three proofs of Minkowski's second inequality in the geometry of numbers (with Woods, Alan; Zassenhaus, Hans) *J. Austral. Math. Soc.* **5** (1965) 453–462.
29. On coverings with convex domains (with Rogers, C. A.; Zassenhaus, H.) *Acta Arith.* **9** (1964) 191–207.
30. A note on the equation $ax^2 - by^2 - cz^2 = 0$ *Indian J. Math.* **4** (1962) 11–12.
31. On the existence of a certain type of basis in a totally real field (with Luthar, Indar S.; Madan, M. L.) *Res. Bull. Panjab Univ.* **12** (1961) 135–137.
32. Some lower bounds on the number of code points in a minimum distance binary code I, II (with Joshi, D. D.; Luthar, Indar S.) *Information and Control* **4** (1961) 313–319, 320–323.
33. Some problems in the geometry of numbers *J. Indian Math. Soc. (N.S.)* **24** (1960) 157–172 (1961).
34. Some transference theorems in the geometry of numbers, *Monatsh. Math.* **62** (1958) 243–249.
35. An analogue of a problem of Mahler, *Res. Bull. Panjab Univ. no.* **109** (1957) 299–302.
36. Maximal covering domains, *Proc. Nat. Inst. Sci. India. Part A* **23** (1957) 540–543.
37. Divided cells, *Res. Bull. Panjab Univ.* **1955** (1955), no. 81, 173–174.
38. Polar reciprocal convex bodies, *Proc. Cambridge Philos. Soc.* **51** (1955) 377–378.
39. An inhomogeneous minimum for non convex star-regions with hexagonal symmetry (with Rogers, K.) *Canad. J. Math.* **7** (1955) 337–346.
40. Four squares and a kth power, *Quart. J. Math., Oxford Ser. (2)* **5** (1954) 191–202.
41. On polar reciprocal convex domains, Addendum. *Proc. Nat. Inst. Sci. India* **20** (1954).
42. Lattice coverings with four-dimensional spheres, *Proc. Cambridge Philos. Soc.* **50** (1954) 203–208.
43. On lattice coverings by spheres, *Proc. Nat. Inst. Sci. India* **20** (1954) 25–52.
44. On polar reciprocal convex domains, *Proc. Nat. Inst. Sci. India* **20** (1954) 119–120.
45. On lattice coverings, *Proc. Nat. Inst. Sci. India* **19** (1953) 447—459.

46. The covering of $n$-dimensional space by spheres (with Davenport, H.) *J. London Math. Soc.* **27** (1952) 224–229.
47. A note on lattice coverings (with Roth, K. F.) *J. Indian Math. Soc. (N.S.)* **16** (1952) 7–12.
48. Covering the planes with convex sets (with Rogers, C. A.) *J. London Math. Soc.* **27** (1952) 304–314.
49. Non-homogeneous binary quadratic forms II, The second minimum of $(x + x_0)^2 - 7(y + y_0)^2$. *Acta Math.* **86** (1951) 31–56.
50. Non-homogeneous binary quadratic forms I, Two theorems of Varnavides. *Acta Math.* **86** (1951) 1–29.
51. Non-homogeneous binary cubic forms, *Proc. Cambridge Philos. Soc.* **47** (1951) 457–460.
52. On the geometry of numbers of non-convex star-regions with hexagonal symmetry, *Philos. Trans. Roy. Soc. London. Ser. A.* **243** (1951) 431–462.
53. On the sign of the Gaussian sum (with Chowla, S.) *Proc. Nat. Inst. Sci. India* **13** (1947) 175–176.
54. The residue of Ramanujan's function $\tau(n)$ to the modulus $2^8$ (with Chowla, S.) *J. London Math. Soc.* **22** (1947) 140–147.
55. On integer roots of the unit matrix (with Chowla, S.) *Proc. Nat. Inst. Sci. India* **13** (1947) 24.
56. On numbers which can be expressed as a sum of two squares (with Chowla, S.) *Proc. Nat. Inst. Sci. India* **13** (1947) 101–103.
57. Congruence properties of Ramanujan's function $\tau(n)$ (with Chowla, S.) *Bull. Amer. Math. Soc.* **53** (1947) 950–955.
58. Congruence properties of Ramanujan's function $\tau(n)$ (with Chowla, S.; Gupta, H.; Lahiri, D. B.) *Quart. J. Math., Oxford Ser.* **18** (1947) 143–146
59. A new congruence property of Ramanujan's function $\tau(n)$ (with Chowla, S.) *Bull. Amer. Math. Soc.* **53** (1947) 768–769.
60. A congruence property of Ramanujan's function $\tau(n)$ (with Chowla, S.; Gupta, H.) *Bull. Amer. Math. Soc.* **53** (1947) 766–767.
61. Ramanujan's function $\tau(n)$—a congruence property. *Bull. Amer. Math. Soc.* **53** (1947) 764–765.
62. A note on Ramanujan's function $\tau(n)$ (with Chowla, S.) *Quart. J. Math., Oxford Ser.* **18** (1947) 122–123.
63. Some new congruence properties of Ramanujan's function $\tau(n)$ (with Chowla, S.) *Math. Student* **14** (1946) 24–26.
64. On integer cube-roots of the unit matrix. *Math. Student* **14** (1946) 69–70 (1948).
65. A congruence property of Ramanujan's function $\tau(n)$ (with Chowla, S.) *Proc. Nat. Inst. Sci. India* **12** (1946) 431–432.
66. On a function of Ramanujan (with Chowla, S.) *Proc. Nat. Inst. Sci. India* **12** (1946) 433.
67. On a function of Ramanujan (with Chowla, S.) *Proc. Nat. Inst. Sci. India* **12** (1946), no. 8, 1 p.

68. A congruence property of Ramanujan's function $\tau(n)$ (with Chowla, S.) *Proc. Nat. Inst. Sci. India* **12** (1946) 431–432.
69. Two congruence properties of Ramanujan's function $\tau(n)$, *J. London Math. Soc.* **21** (1946) 91–93.
70. On complete primitive residue sets, *Bull. Calcutta Math. Soc.* **38** (1946) 113–116.
71. On integer roots of the unit matrix (with Chowla, S.) *Science and Culture* **12** (1946) 105.

PANJAB UNIVERSITY
CHANDIGARH
PANJAB UNIVERSITY, CHANDIGARH
INTERNATIONAL CONFERENCE ON NUMBER THEORY AND DISCRETE GEOMETRY
NOV.-30 TO DEC.-3, 2005

# Contents

Proc. Int. Conf. – Number Theory and Discrete Geometry No. 6, 2008, pp. 1–12.

# A Problem on the Fractional Parts of the Powers of 3/2 and Related Questions

Sukumar Das Adhikari and Purusottam Rath

Harish-Chandra Research Institute, (Formerly, Mehta Research Institute),
Chhatnag Road, Jhusi, Allahabad 211 019, India
e-mail: adhikari@mri.ernet.in, rath@mri.ernet.in

**Abstract.** For a real number $x$, denoting the fractional part of $x$ by $\{x\}$, we here discuss on the distribution of the sequence $\{(3/2)^n\}$ and some related questions. After a discussion on the early results, we also have a glimpse of the recent progress in this area.

*Dedicated to Prof. R. P. Bambah on his 80th birthday*

## 1. Introduction

For a real number $x$, denoting the fractional part of $x$ by $\{x\}$ (that is, $\{x\} = x - [x]$, where $[x]$ is the largest integer $\leq x$), the connection between the sequence $\{(3/2)^n\}$ and Waring's problem is well known (see [20], for example).

In general, for real numbers $\xi > 0, \theta > 1$, the study of the distribution of the sequence $\{\xi\theta^n\}_{n\geq 0}$, is one of the intriguing problems in number theory and has a long history.

For a fixed $\theta$, it is known (Weyl [44]) that the sequence $\{\xi\theta^n\}$ is uniformly distributed in $[0, 1]$ for almost all $\xi$.

Similarly, it is also well known (Koksma [23]) that for a fixed $\xi$, the sequence $\{\xi\theta^n\}$ is uniformly distributed in $[0, 1]$ for almost all real $\theta > 1$.

However, for very simple particular sequences like $\left(\frac{3}{2}\right)^n$ or $e^n$, it is not yet known whether they are uniformly distributed modulo 1 or even whether they are everywhere dense modulo 1 in $[0, 1]$.

In the present expository article, in Section 2, we shall discuss some of the results which appeared after the early results of Weyl [44] and Koksma [23] and shall cover the developments during the period 1935–1980. In Section 3, we shall try to have a glimpse of the recent progress in this area.

*Notations.* Apart from the notations $[x]$ and $\{x\}$ introduced at the beginning of the introduction, for a real number $x$ we shall use $\|x\|$ to denote the absolute value of the difference between $x$ and the nearest integer. The symbols $\mathbb{Z}$, $\mathbb{Q}$, $\mathbb{R}$ will respectively denote the integers, the rationals and the reals. Other notations used in this article will be defined at appropriate places.

## 2. Developments during the period 1935–1980

For real numbers $\xi > 0, \theta > 1$, let us denote the set of limit points of the sequence $\{\xi\theta^n\}_{n\geq 0}$ by $G(\xi, \theta)$.

Following an argument of A. Weil (see Vijayaraghavan [41]), we present an elementary proof of a result of Vijayaraghavan; one of the early non-trivial results on the set $G(\xi, p/q)$, where $p$ and $q$ are integers with $(p, q) = 1$, $p > q > 1$. Actually, Vijayaraghavan was dealing with the case $\xi = 1$; however, the argument of A. Weil goes through for all real numbers $\xi > 0$. A more general result of this nature will be mentioned later (see the last statement of Theorem 2.11) in this section, after we introduce the notion of PV numbers.

**Theorem 2.1.** *For any real $\xi > 0$ and any rational $p/q$, where the integers $p, q$ satisfy $(p, q) = 1$, $p > q > 1$, the set $G(\xi, p/q)$ is infinite.*

*Proof.* Assume the contrary and let $G(\xi, p/q) = \{x_1, x_2, \ldots, x_k\}$.

We write $\xi(p/q)^n$ as $\xi(p/q)^n = e_n + f_n$, where $e_n$ is the integral part and $f_n = \{\xi(p/q)^n\}$ is the fractional part of $\xi(p/q)^n$.

If all the limit points $x_1, x_2, \ldots, x_k$ are rational, then there exists an integer $M$ such that $Mx_1, \ldots, Mx_k$ are all integers. We have

$$M\xi(p/q)^n = Me_n + Mf_n = a_n + \delta_n$$

where $a_n$ is the integer nearest to $M\xi(p/q)^n$. Thus $|\delta_n| \leq 1/2$.

Since $\delta_n = Me_n + Mf_n - a_n$, every limit point of $\delta_n$ is an integer and hence $\delta_n \to 0$ as $n \to \infty$.

Clearly, for all suitably large values of $n$, we have

$$\delta_n \neq 0, \qquad \left|\delta_{n+1} - \frac{p}{q}\delta_n\right| < 1/q.$$

However, $\left|\delta_{n+1} - \frac{p}{q}\delta_n\right| = \left|\frac{p}{q}a_n - a_{n+1}\right| \neq 0$ would imply that

$$\left|\delta_{n+1} - \frac{p}{q}\delta_n\right| \geq 1/q.$$

Hence for all suitably large values of $n$, $\delta_{n+1} = \frac{p}{q}\delta_n$ which contradicts the facts that $\delta_n \neq 0$ and $\delta_n \to 0$ as $n \to \infty$. Thus $G(\xi, p/q)$ can not be finite if all the limit points are rational.

Now, suppose at least one of the limit points, say $x_1$, is irrational. Let $f_{n_r} \to x_1$ as $r \to \infty$.

Let $m$ be a fixed integer and let $\lambda_r = n_r + m$. Then

$$f_{\lambda_r} = \xi\left(\frac{p}{q}\right)^{n_r+m} - e_{n_r+m} = \left(\frac{p}{q}\right)^m (e_{n_r} + f_{n_r}) - e_{n_r+m}.$$

Thus we have,

$$f_{\lambda_r} \equiv (p/q)^m f_{n_r} \pmod{1/q^m}.$$

Now the sequence $f_{\lambda_r}$ will have at least one limit point say $\eta_m$ and since $f_{n_r} \to x_1$ as $r \to \infty$, it follows that

$$\eta_m \equiv (p/q)^m x_1 \pmod{1/q^m}$$

Since $x_1$ is irrational, we see from above that all such $\eta_m$' s are distinct for distinct integers $m$. Thus $G(\xi, p/q)$ is infinite.

From the above theorem, we see that for $\theta = p/q$, where $p$ and $q$ are integers, $(p, q) = 1$, $p > q > 1$, the set $\{\theta^n\}_{n=1}^{\infty}$ (and hence in particular, the set $\left\{\left(\frac{3}{2}\right)^n\right\}_{n=1}^{\infty}$) has infinitely many limit points. But, it is not yet known whether the interval [0, 1/2) (or [1/2, 1)) contains infinitely many limit points of the sequence $\{(3/2)^n\}$.

Defining a *Z-number* to be a real number $\alpha > 0$ satisfying the condition that $0 \leq \{\alpha(3/2)^n\} < \frac{1}{2}$ for all $n \geq 0$, Mahler [27] proved that the set of $Z$-numbers is at most countable.

More precisely, Mahler proved that:

**Theorem 2.2.** *Given any integer $g \geq 0$, there exists at most one $Z$-number in $[g, g+1)$.*

For $x > 0$, if $Z(x)$ denotes the set of $Z$-numbers $\leq x$, and $|Z(x)|$ the cardinality of $Z(x)$, then Mahler also established that:

**Theorem 2.3.**

$$|Z(x)| = O(x^{0.7}).$$

In the other direction, as a consequence of a more general result of Boyd [5], it follows that:

**Theorem 2.4.** *In any interval $[a, b]$ with $3 < a < b$, there are uncountably many numbers $\beta$ and for each $\beta$ infinitely many $\eta = \eta(\beta)$ such that*

$$\|\eta\beta^n\| \leq \frac{1}{(a-1)(\beta-1)} \quad \textit{for } n = 0, 1, 2, \ldots$$

The following result of Tijdeman [40] includes the cases $\beta > 2$.

**Theorem 2.5.** *If $\beta$ is a real number with $\beta > 2$, then for every integer $g \geq 0$, there exists a real number $\eta \in (g, g+1)$ such that*

$$\{\eta\beta^n\} \in \left[0, \frac{1}{\beta - 1}\right].$$

*Further, if $\beta = (2k+1)/2$ for some integer $k > 1$, then in the above statement, the interval $\left[0, \frac{1}{\beta-1}\right]$ can be replaced by $\left[0, \frac{1}{2(\beta-1)}\right]$.*

The question of existence of a $Z$-number is still open and is closely related (see [27], [25]) to the $3x + 1$ problem.

The notion of $Z$-numbers was generalised by Flatto [17] to any $0 < t < 1$ and $\theta > 1$ by defining

$$Z_{t,\theta} = \{\alpha > 0 : 0 \leq \{\alpha\theta^n\} < t, \forall n \geq 0\}.$$

We shall report on the results of the above paper of Flatto in the next section.

Elements of $Z_{\frac{1}{2},\frac{3}{2}}$ are the $Z$-numbers defined by Mahler.

With the above notation, the result of Tijdeman [40], mentioned above, says that for $\theta = (2k+1)/2$, $k > 1$ and $t = 1/(2k-1)$, there exists at least one element of $Z_{t,\theta}$ in $[g, g+1)$ for every integer $g \geq 0$.

A result of Pollington [31], which also had appeared before the above mentioned paper of Flatto, says that for any integer $g \geq 0$, and for integers $p, q$ satisfying $(p, q) = 1$, $p > q > 1$, there are no nonzero numbers $\alpha$ with $\{\alpha(p/q)^n\} < \frac{1}{p}$, for all $n \geq 0$; in other words, $Z_{\frac{1}{p},\frac{p}{q}}$ is empty.

We now state another result of Pollington after providing some background to the problem.

It had been established by Erdős and Taylor [16] that:

**Theorem 2.6.** *If* $(n_k)$ *is an increasing sequence of positive integers such that* $t_k = n_{k+1}/n_k > \rho > 1$, *then the set* $E$ *of non-zero real numbers* $x$ *such that* $\{n_k x\}$ *is not uniformly distributed in* $(0, 1)$ *has Hausdorff dimension* 1 *and hence is uncountable.*

Later, Erdős [15] asked the following question:

Given an increasing sequence of positive integers $(n_k)$ such that $t_k = n_{k+1}/n_k > \rho > 1$, is it true that there always exists an irrational $\xi$ for which the sequence $\{n_k\xi\}$ is not everywhere dense?

Strzelecki [39] showed that:

**Theorem 2.7.** *If* $\rho > 5^{1/3}$, *and* $(x_k)$ *is an increasing sequence of positive real numbers, not necessarily integers, such that* $t_k = x_{k+1}/x_k > \rho$, *then there exists a non-zero real number* $\xi$ *such that* $\{x_k\xi\} \in [\beta, 1-\beta]$ *for* $k = 1, 2, \ldots$, *for some* $\beta > 0$.

The following corollary of a result of Pollington [32] provides a complete answer to the above mentioned question of Erdős.

**Theorem 2.8.** *If* $(x_k)$ *is an increasing sequence of positive real numbers, such that* $t_k = x_{k+1}/x_k > \rho > 1$, *then the set of real numbers* $\xi$ *such that* $\{x_k\xi\}$ *is not dense in the unit interval has Hausdorff dimension 1.*

A corollary to yet another result of Pollington [31] [33] says that:

**Theorem 2.9.** *Given any real number* $\xi > 0$ *and* $0 < \delta < 1$, *the set of real numbers* $\theta$ *for which*

$$\{\xi\theta^n\} \in [0, \delta], \quad \textit{for all } n = 1, 2, 3, \ldots$$

*has Hausdorff dimension 1.*

Improving upon a related result of Choquet [10], corresponding to the particular sequence $\{\xi(3/2)^n\}$, Pollington [34] proved that:

**Theorem 2.10.** $E(0.108)$ *is infinite and countable whereas* $E(0.096)$ *is uncountable, where given* $\lambda > 0$,

$$E(\lambda) = \{\xi \in \mathbb{R} : \|\xi(3/2)^n\| > \lambda, \quad \textit{for all integers } n \geq 0\}.$$

We now briefly mention about the study of numbers (other than rational integers) $\theta > 1$, such that the sequence $\theta^n$ is rather badly distributed modulo 1. Considering the algebraic integer $\omega = \frac{1}{2}(1 + \sqrt{5})$, it is not difficult to observe that, the sequence $\{\omega^n\}_{n=1}^{\infty}$ has a single limit point modulo 1. Indeed, writing $\omega' = \frac{1}{2}(1 - \sqrt{5})$, $\omega^n + \omega'^n$ is a rational integer and therefore observing that $\omega'^n \to 0$ as $n \to \infty$, we have $\omega^n \to 0$ (modulo 1). In fact, the number $\frac{1}{2}(1 + \sqrt{5})$ is one among a particular class of numbers known for such behaviour. An algebraic integer $\alpha > 1$ is called a *Pisot number* (or a *Pisot-Vijayaraghavan number* or a *PV number*) if the moduli of all other (if any) conjugates of $\alpha$ are strictly smaller than 1. It was proved by Pisot [28] (presented also in [35], [4]) that in every real algebraic field, there exist PV numbers having the degree of the field. Vijayaraghavan [42] also observed that there are algebraic integers of all degrees among the PV numbers.

It is known [28] [29] [42] (see also [35]) that

**Theorem 2.11.** *If $\theta$ is a PV number, then for any algebraic integer $\lambda \in \mathbb{Q}(\theta)$, $\|\lambda\theta^n\| \to 0$ as $n \to \infty$.*

*Also, if $\theta > 1$ is an algebraic number such that there exists a real number $\lambda$ with the property $\|\lambda\theta^n\| \to 0$ as $n \to \infty$, then $\theta$ is a Pisot number and $\lambda \in \mathbb{Q}(\theta)$. More generally, if $\theta > 1$ is an algebraic number, such that $G(\lambda, \theta)$ is finite, then $\theta$ is a PV number and $\lambda \in \mathbb{Q}(\theta)$.*

We should remark that the above characterization of a real algebraic number $\theta > 1$ with the property that there exists a real number $\lambda > 0$ with the property $\|\lambda\theta^n\| \to 0$ as $n \to \infty$ (namely that such a $\theta$ is then necessarily an algebraic integer such that moduli of all other conjugates of $\theta$ are strictly smaller than 1), had been found by Hardy [19], much before Pisot and Vijayaraghavan embarked upon the study of these numbers. Regarding the general problem to characterize all real $\theta > 1$ with the above property, we would like to quote Hardy [19]:

"(It) appears to be one of considerable interest and difficulty".

Indeed, after more than ninety years since the above remark was made by Hardy, with our present knowledge, one can neither find nor rule out the possibility of existence of a transcendental number $\theta$ with the property $\|\lambda\theta^n\| \to 0$, as $n \to \infty$, for some $\lambda > 0$; however, the set of all $\theta > 1$, having this property is known [43] [35] [28] to be countable. In fact, it is known [29] that:

**Theorem 2.12.** *The set of all pairs $(\lambda, \theta)$ with $\theta > 1$ and $\lambda$ real, such that the set $G(\lambda, \theta)$ consists of finitely many points is countable.*

In the same paper [29], Pisot also proved the following result.

**Theorem 2.13.** *If $\xi \geq 1$ and $\theta > 1$ are real numbers such that*

$$\|\xi\theta^n\| \leq \epsilon \text{ for all integers } n \geq 0,$$

*where*

$$\epsilon = \frac{1}{2e\theta(\theta + 1)(1 + \log \xi)}$$

*then $\theta$ is a Pisot number or a Salem number, deg $\theta \leq [\log \xi] + 1$ and $\xi \in \mathbb{Q}(\theta)$.*

For some later developments in this direction we refer to the papers of Decomps-Guilloux and Grandet-Hugot [12] and Hata [21].

If the condition $\|\xi\theta^n\| \le \epsilon$ in the above result is replaced by $\|\xi\theta^n\| \le 10\epsilon$, then a striking fact which follows from a result of Boyd [5] is that $\theta$ need not even be algebraic.

A *Salem number* (introduced by Salem in [36]) is an algebraic integer $\tau > 1$ having all its other conjugates in the unit disc with at least two lying on the unit circle. If $P(x) \in \mathbb{Z}[x]$ is the minimal polynomial for a Salem number $\tau$, then $P(x)$ will have two roots $\alpha$ and $1/\alpha$ on the unit circle, ensuring that $P(x)$ is a reciprocal polynomial. Therefore, besides $\tau$ and $1/\tau$, all other conjugates of $\tau$ will have modulus 1. Since there will be an even number of imaginary roots of modulus 1, the degree of $P$ has to be even and at least 4.

For a Salem number $\tau$, although the sequence $\{\tau^n\}$ is dense in $[0, 1]$, it is not uniformly distributed [30] [35]. However, recently it has been shown by Akiyama and Tanigawa [3] that if the degree of $\tau$ is large, then $\{\tau^n\}$ is 'almost uniformly distributed' in $[0, 1]$.

We mention that before the introduction of Salem numbers, Vijayaraghavan [42] had already considered the larger positive root $\theta$ of the polynomial $f(x) = x^4 - x^3 - x^2 - x + 1$ and had showed that $\theta$ is an example of an algebraic integer such that $\{\theta^n\}$ is dense in $[0, 1]$; the polynomial $f(x)$ being a reciprocal polynomial having two real positive roots and two non-real roots, it is clear that $\theta$ is a Salem number.

The set of Pisot numbers is closed [36] and every Pisot number is a limit point of the set of Salem numbers [37]. It is not known whether there are other limit points of the set of Salem numbers.

In the next section we shall mention a recent result of Corvaja and Zannier [11] related to PV numbers. For further information regarding Pisot and Salem numbers, we refer to [35], [8] and [4].

## 3. Developments after 1980

With an idea from the paper of Mahler [27] mentioned earlier, taken as an essential ingredient for their proof, in a striking progress, Flatto, Lagarias and Pollington [18] have shown that:

**Theorem 3.1.** *For integers $p > q \ge 2$ with gcd $(p, q) = 1$,*

$$\limsup_{n\to\infty}\{\xi(p/q)^n\} - \liminf_{n\to\infty}\{\xi(p/q)^n\} \ge \frac{1}{p}, \tag{1}$$

*holds for all real $\xi > 0$.*

From the above result it follows that for integers $p > q \ge 2$ with gcd $(p, q) = 1$, and any real $\xi > 0$, $\{\xi(p/q)^n\}$ can not be contained in any interval of length $1/(p+\epsilon)$ for any $\epsilon > 0$.

For $p/q = 3/2$, it is expected that there exists no real $\xi$ for which $\{\xi(3/2)^n\} \in [0, 1/2)$ for all $n$. In other words, using the terminology of Z-numbers, as defined before the statement of Theorem 2.2, it is expected that Z-numbers do not exist.

Therefore, it will be interesting to have the following result

$$\limsup_{n\to\infty}\{(3/2)^n\} - \liminf_{n\to\infty}\{(3/2)^n\} > \frac{1}{2},$$

which would have implied that $\{(3/2)^n\}$ has limit points in both of the intervals $[0, 1/2)$ and $[1/2, 1)$.

For an interesting recent result regarding positive numbers $z$ such that $\{z(3/2)^n\} \in [0, 1/3[\cup[2/3, 1[$, for all $n = 0, 1, 2, \ldots$, we refer to a paper of Akiyama, Frougny and Sakarovitch (Corollary 4, [2]).

In Flatto's article [17], apart from the results of Mahler (Theorem 2.2), Tijdeman (Theorem 2.5) and Pollington (and some generalizations of those results), mentioned in the last section, some new results were established; for instance, it was proved that if $p > q^2$ and $\frac{q-1}{p-q} < t < \frac{q}{p}$, there is precisely one element of $Z_{t,\frac{p}{q}}$ in each integer interval. Flatto [17] has also improved Theorem 2.3 of Mahler by replacing the number 0.7 by $\log_2(3/2) = 0.59\ldots$.

For further generalizations of the concept of $Z$-numbers and some interesting results, we refer to the above mentioned paper of Flatto, Lagarias and Pollington [18] and a recent paper of Bugeaud [6].

We are now going to state an important recent result of Dubickas [13] which generalizes the result (1) of Flatto, Lagarias and Pollington [18].

We need to introduce some terminology.

For a polynomial $f$ over the field of complex numbers, its *length* $L(f)$ is defined as the sum of the moduli of its coefficients. If $\alpha$ is an algebraic number, then its *length* is defined to be the length of its minimal polynomial in $\mathbf{Z}[x]$. Thus, if the minimal polynomial of $\alpha$ in $\mathbb{Z}[x]$ is $P(x) = a_d x^d + \cdots + a_0$, then its length $L(\alpha)$ is given by

$$L(\alpha) = L(P) = \sum_{j=0}^{d} |a_j|.$$

For an algebraic number $\alpha$, its *reduced length* $l(\alpha)$ (and $l(P)$ for the corresponding minimal polynomial $P$) is defined by

$$l(\alpha) = l(P) = \inf_{G\in\Gamma} L(PG),$$

where

$$\Gamma = \{b_0 + b_1 x + \cdots + b_m x_m \in \mathbb{R}[x],\ \text{where } b_0 = 1 \text{ or } b_m = 1 \mid m \geq 0\}.$$

If for an algebraic number $\alpha$, the moduli of $\alpha$ and its conjugates are all strictly greater (smaller) than 1, then it can be shown that (see [13])

$$l(\alpha) = \max\{a_d, |a_0|\}, \tag{2}$$

where $P(x) = a_d x^d + \cdots + a_0$, with $a_d > 0$, is the minimal polynomial of $\alpha$ in $\mathbb{Z}[x]$.

Dubickas [13] has proved the following:

**Theorem 3.2.** *If $\alpha > 1$ is an arbitrary algebraic number and $\xi$ is an arbitrary positive number, such that $\xi$ lies outside the field $\mathbb{Q}(\alpha)$ if $\alpha$ is a Pisot number or a Salem number, then*

$$\limsup_{n\to\infty}\{\xi\alpha^n\} - \liminf_{n\to\infty}\{\xi\alpha^n\} \geq \frac{1}{l(\alpha)}. \tag{3}$$

For integers $p, q$ with $p > q \geq 1$ and gcd $(p, q) = 1$, since $p/q$ is the root of $p - qx$, from (2) we have $l(p/q) = \max\{p, q\} = p$ and hence from (3), one recovers the result (1) of Flatto, Lagarias and Pollington (along with the additional case $q = 1$).

Recently, Zaïmi [45] has shown that if $\alpha$ is a Salem number and $\xi$ is a non-zero element of the field $\mathbb{Q}(\alpha)$, then the inequality (3) remains true only if $\alpha - 1$ is a unit.

Before taking up the theme of distribution functions, we shall mention some of the other important results related to fractional parts of powers published very recently.

An old result of Mahler [26] says that if $\alpha > 1$ is rational but not an integer and $0 < l < 1$, then $\|\alpha^n\| > l^n$ except for a finite set of integers $n$ depending on $\alpha$ and $l$.

Answering a question of Mahler, Corvaja and Zannier [11] have recently proved the following:

**Theorem 3.3.** *Let $\alpha > 1$ be a real algebraic number and let $0 < l < 1$. Suppose that $\|\alpha^n\| < l^n$ for infinitely many natural numbers $n$. Then there is a positive integer $d$ such that $\alpha^d$ is a Pisot number. In particular, $\alpha$ is an algebraic integer.*

It is interesting to note that the conclusion is not generally true without the assumption that $\alpha$ is algebraic. Indeed, Corvaja and Zannier [11] have shown that there exists a real number $\alpha > 1$ such that $\|\alpha^n\| < 2^{-n}$ for infinitely many natural numbers $n$, but $\alpha^d$ is not a Pisot number for any positive integer $d$ (and hence not algebraic by the above theorem).

The next result related to Mahler's problem follows as a corollary to a recent result of Dubickas [14].

**Theorem 3.4.** *For any $\xi \neq 0$, the interval $[0.238\ldots, 0.762\ldots]$ contains a limit point of the sequence $\{\xi(3/2)^n\}$, $n = 1, 2, \ldots$.*

Another important recent result of Bugeaud and Dubickas [7] describes all irrational $\xi > 0$ with the property that the fractional parts $\{\xi b^n\}, n \geq 0$, all belong to a semi-open or an open interval of length $1/b$.

For the remaining portion of our article, we now briefly explain the concept of distribution functions of a sequence; we refer to [24] for details.

A *distribution function* $g(x)$ is a real valued, non-decreasing function defined on $[0, 1]$ for which $g(0) = 0$ and $g(1) = 1$.

Let $\Delta = \{x_n\}_{n=1}^{\infty}$ be a sequence with $x_n \in [0, 1)$. For any positive integer $N$ and a subinterval $I \subset [0, 1]$, we define the following counting function:

$$A(I; N; \Delta) = \#\{x_n \mid 1 \leq n \leq N, x_n \in I\}.$$

A distribution function $g$ is called *a distribution function of the sequence* $\Delta$ if there exists an increasing sequence of positive integers $N_1, N_2, \ldots$ such that

$$\lim_{k\to\infty} \frac{A([0,x); N_k; \Delta)}{N_k} = g(x), \text{ for every } x \in [0,1].$$

The sequence $\Delta$ is said to have the *asymptotic distribution function* $g$ if

$$\lim_{k\to\infty} \frac{A([0,x); k; \Delta)}{k} = g(x), \text{ for every } x \in [0,1].$$

In terms of distribution functions, the following statement clearly implies non-existence of $Z$-numbers:

*If $g(x)$ is a distribution function of $\{\xi(3/2)^n\}$ such that $g(x)$ is constant for all $x$ in an interval $I \subset [0,1]$, then $|I| < \frac{1}{2}$.*

Thus the study of distribution functions of $\{\xi(p/q)^n\}$ becomes relevant.

For a distribution function $g$ of a given sequence, a set $X \subset [0,1]$ is said to be a *set of uniqueness of* $g$, if defining $g$ on $X$, uniquely determines it in all of $[0,1]$. Stating differently, if $g_1$ and $g_2$ are any two distribution functions of a sequence which agree on $X$, then they agree on [0,1]. Strauch [38] proved the following theorem related to distribution functions of the sequence $\{\xi(3/2)^n\}$.

**Theorem 3.5.** *Let*

$$I_1 = [0,1/3], I_2 = [1/3,2/3], I_3 = [2/3,1].$$

*Then for any $i, j$ with $1 \le i \ne j \le 3$, the set $X = I_i \cup I_j$ is a set of uniqueness of any distribution function $g$ of $\{\xi(3/2)^n\}$.*

In a very recent paper of Adhikari, Rath and Saradha [1], Theorem 3.5 has been extended:

**Theorem 3.6.** *Suppose $g$ is a distribution function of $\{\xi(p/q)^n\}$ where $p > q > 1$ are positive integers with* $\gcd(p,q) = 1$. *Let*

$$I_i = \left(\frac{i-1}{p}, \frac{i}{p}\right) \text{ and } J_i = [0,1] - I_i \text{ for } 1 \le i \le p.$$

*Further, we assume that $p \ge q^2 - q$ if $j/q \in I_i$ for some $j$ with $1 \le j < q$. Then $X = J_i$ for $1 \le i \le p$ is a set of uniqueness of $g$.*

Another theorem proved in [1] is the following:

**Theorem 3.7.** *Suppose $g$ is a distribution function of $\{\xi(p/q)^n\}$ with $p, q$ as in Theorem 3.6. Then any interval $\left[a, a + \frac{p-1}{p}\right] \subset [0,1]$ of length $\frac{p-1}{p}$ is a set of uniqueness of $g$.*

We observe that, restricting to the case $p = 3, q = 2$, the above theorem describes a different class of sets of uniqueness of distribution functions of $\{\xi(3/2)^n\}$ which is not covered by Theorem 3.5. of Strauch.

Determining the existence of asymptotic distribution function of sequences of the form $\{\xi\theta^n\}$ is rather difficult. The result of Weyl [44] mentioned in the introduction says that for almost all $\xi$, $\{\xi\theta^n\}$ is uniformly distributed in $[0, 1]$ and hence this sequence has the asymptotic distribution function $g(x) = x$.

In the other direction, Helson and Kahane [22] established the existence of uncountably many $\xi$ such that $\{\xi\theta^n\}$ does not have an asymptotic distribution function where $\theta > 1$ is any fixed real number. Therefore, for positive integers $p, q$ as in Theorem 3.6, considering the sequence $\{\xi(p/q)^n\}$, for uncountably many $\xi$, the sequence does not have an asymptotic distribution function and hence is not uniformly distributed. However, for each such $\xi$, Theorem 3.6 (with notations as in the theorem), rules out the possibility that all but finitely many elements of the sequence $\{\xi(p/q)^n\}$ are in a single interval $I_i$, for some fixed $i$, $1 \leq i \leq p$. Since otherwise, any distribution function (there exists at least one, by Helly selection principle; see [24], Theorem 7.1, for instance) $g(x)$ of such $\{\xi(p/q)^n\}$ satisfies

$$g(x) = \begin{cases} 0 \text{ for } x \in [0, (i-1)/p] \\ 1 \text{ for } x \in [i/p, 1] \end{cases}$$

and therefore by Theorem 3.6, $\{\xi(p/q)^n\}$ will then have exactly one distribution function which will have to be its asymptotic distribution function (see [24], for instance).

As an application of Theorem 3.5, Strauch (see Section 5 of [38]) had showed that the distribution function $g$ given by

$$g(x) = \begin{cases} x \text{ for } x \in [0, 2/3] \\ x^2 - 2/3x + 2/3 \text{ for } x \in [2/3, 1] \end{cases}$$

is not a distribution function of $\{\xi(3/2)^n\}$ for any $\xi$.

In the spirit of the example given by Strauch, it has been shown in [1] that distribution function

$$g_1(x) = \begin{cases} x \text{ for } x \in [0, (p-1)/p] \\ h(x) \text{ for } x \in [(p-1)/p, 1], \end{cases}$$

where $h : [(p-1)/p, 1] \to [(p-1)/p, 1]$ is any non-decreasing function other than the identity map with $h(\frac{p-1}{p}) = \frac{p-1}{p}$ and $h(1) = 1$, cannot be a distribution function for the sequence $\{\xi(p/q)^n\}$ for any $\xi > 0$.

A conjecture of Strauch [38], says that every measurable set $X \subset [0, 1]$ having measure at least $2/3$ is a set of uniqueness of any distribution function of $\{\xi(3/2)^n\}$ for any $\xi > 0$. Strauch also showed that each of the sets $Y = [2/9, 1/3] \cup [1/2, 1]$ and $Z = [0, 1/2] \cup [2/3, 7/9]$ is a set of uniqueness of any such distribution function of $\{\xi(3/2)^n\}$; one observes that both $Y$ and $Z$ are of measure $11/18 < 2/3$. In light of Theorem 3.7, it has been asked in [1] whether there exists an interval $I$ of measure less than $2/3$ such that $I$ is a set of uniqueness of any distribution function of $\{\xi(3/2)^n\}$.

## Acknowledgement

We would like to thank Yann Bugeaud for going through our manuscript and making some helpful remarks.

## References


[1] S. D. Adhikari, P. Rath and N. Saradha, On the sets of uniqueness of a distribution function of $\{\xi(p/q)^n\}$, *Acta Arith.* **119** (2005) 307–316

[2] S. Akiyama, C. Frougny and J. Sakarovitch, Powers of rationals modulo 1 and rational base number systems, Preprint

[3] S. Akiyama and Y. Tanigawa, Salem numbers and uniform distribution modulo, *I. Publ. Math. Debrecen* **64** (2004) no. 3–4 329–341

[4] M. J. Bertin, A. Decomps-Guilloux, M. Grandet-Hugot, M. Pathiaux-Delefosse and J. P. Schreiber, Pisot and Salem Numbers, Birkhäuser Verlag, (1992)

[5] D. W. Boyd, Transcendental numbers with badly distributed powers, *Proc. Amer. Math. Soc.* **23** (1969) 424–427

[6] Y. Bugeaud, Linear mod one transformations and the distribution of fractional parts $\{\xi(p/q)^n\}$, *Acta Arith.* **114** (2004) 301–311

[7] Y. Bugeaud and A. Dubickas, Fractional parts of powers and Sturmian words, *C.R. Acad. Sci. Paris, Sér. I* **341** (2005) 69–74

[8] D. G. Cantor, On power series with only finitely many coefficients mod 1: solution of a problem of Pisot and Salem, *Acta Arith.* **34** (1977) 43–55

[9] J. W. S. Cassels, An Introduction to Diophantine Approximation, *Cambridge Univ. Press*, London, (1957)

[10] G. Choquet, Construction effective de suites $\{k(3/2)^n\}$. Étude des mesures $(3/2)$-stables, *C. R. Acad. Sci. Paris Sér. A-B* **291** (1980) A69–A74

[11] P. Corvaja and U. Zannier, On the rational approximations to the powers of an algebraic number: Solution of two problems of Mahler and Mendès France, *Acta Math.* **193** (2004) 175–191

[12] A. Decomps-Guilloux and M. Grandet-Hugot, Nouvelles caractérisations des nombres de Pisot et de Salem, *Acta Arith.* **50** (1988) no. 2 155–170

[13] A. Dubickas, Arithmetical properties of powers of algebraic numbers, *Bull. London Math. Soc.* **38** (2006) no. 1 70–80

[14] A. Dubickas, On the distance from a rational power to the nearest integer, *J. Number Theory* **117** (2006) no. 1 222–239

[15] P. Erdős, Problems and results in Diophantine approximations II, Repartition modulo 1, *Lect. notes in Math.* **475** Springer Verlag, (1975)

[16] P. Erdős and S. J. Taylor, On the set of points of convergence of a lacunary trigonometric series and the equidistribution properties of related sequences, *Proc. London Math. Soc.* (3) **7** (1957) 598–615

[17] L. Flatto, $Z$-numbers and $\beta$-transformations, Symbolic dynamics and its applications (New Haven, CT, 1991), *Contemp. Math.* **135** (1992) 181–201

[18] L. Flatto, J. C. Lagarias and A. D. Pollington, On the range of fractional parts $\{\xi(p/q)^n\}$, *Acta Arith.* **70** (1995) 125–147

[19] G. H. Hardy, A problem of diophantine approximation, *J. Indian Math. Soc.* **11** (1919) 205–243

[20] G. H. Hardy and E. M. Wright, An introduction to the Theory of numbers, 5th ed., Oxford Univ. Press, London, (1979)

[21] M. Hata, Note on the fractional parts of $\lambda\theta^n$, *Acta Arith.* **120** (2005) no. 2 153–157

[22] H. Helson and J. -P. Kahane, A Fourier method in diophantine problems, *J. Analyse Math.* **15** (1965) 245–262

[23] J. F. Koksma, Ein mengentheoretischer satz über die gleichverteilung modulo eins, *Compositio Math.* **2** (1935) 250–258

[24] L. Kuipers and H. Niederreiter, Uniform Distribution of Sequences, Wiley, New York, 1974

[25] J. C. Lagarias, The $3x+1$ problem and its generalizations , *Amer. Math.* Monthly **92** (1985) 3–23; Organic Mathematics (Burnaby, BC, 1995), CMS *Conf. Proc., 20, Amer. Math. Soc.* Providence, (1997) 305–334

[26] K. Mahler, On the fractional parts of the powers of a rational number, II, *Mathematika* **4** (1957) 122–124

[27] K. Mahler, An unsolved problem on the powers of 3/2, *J. Austral. Math. Soc.* **8** (1968) 313–321

[28] C. Pisot, La répartition modulo 1 et les nombres algébriques, *Ann. Scu. Norm. Sup. Pisa* **7** (1938) 205–248

[29] C. Pisot, Répartition (mod. 1) des puissances successives des nombres réels, *Comment. Math. Helv.* **19** (1946) 153–159

[30] C. Pisot and R. Salem, Distributation modulo 1 of the powers of real numbers larger than 1, *Comp. Math.* **16** (1964) 164–168

[31] A. D. Pollington, Interval constructions in the theory of numbers, Ph.D Thesis, Univ. London, London, 1978

[32] A. D. Pollington, On the density of sequence $\{n_k\xi\}$, *Illinois Journal of Mathematics* **23** no. 4 (1979) 511–515

[33] A. D. Pollington, The Hausdorff dimension of certain sets related to sequences which are not dense mod 1, *Quart. J. Math.* Oxford (2) **31** (1980) 351–361

[34] A. D. Pollington, Sor les suites $\{k\theta^n\}$, *C.R. Acad. Sci. Paris, Sér. I* **296** no. 23 (1983) 941–943

[35] R. Salem, Algebraic Numbers and Fourier Analysis, *Heath Math. Monographs*, Boston, Mass. 1963

[36] R. Salem, A remarkable class of algebraic integers. Proof of a conjecture of Vijayaraghavan, *Duke Math. J.* **11** (1944) 103–108

[37] R. Salem, Power series with integral coefficients, *Duke Math. J.* **12** (1945) 153–172

[38] O. Strauch, Distribution functions of $\xi(3/2)^n$ mod 1, *Acta Arith.* **81** (1997) 25–35

[39] E. Strzelecki, On sequences $\{\xi t_n(\mathrm{mod}\,1)\}$, *Canad. Math. Bull.* **18** (1975) 727–738

[40] R. Tijdeman, Note on Mahler's 3/2-problem, *Norske Vid. Selsk. Skr.* **16** (1972) 1–4

[41] T. Vijayaraghavan, On the fractional parts of the powers of a number, *I J. London Math. Soc.* **15** (1940) 159–160

[42] T. Vijayaraghavan, On the fractional parts of the powers of a number, *II Proc. Camb. Phil. Soc.* **37** (1941) 349–357

[43] T. Vijayaraghavan, On the fractional parts of the powers of a number, *III J. London Math. Soc.* **17** (1942) 137–138

[44] H. Weyl, Über die gleichverteilung von zalen modulo eins, *Math. Ann.* **77** (1916) 313–352

[45] T. Zaïmi, An arithmetical property of powers of Salem numbers, *J. Number Theory* **120** (2006) no. 1 179–191

Proc. Int. Conf. – Number Theory and Discrete Geometry No. 6, 2008, pp. 13–18.

# Idempotent Generators of Irreducible Cyclic Codes

Gurmeet K. Bakshi, Madhu Raka and Anuradha Sharma*

Centre for Advanced Study in Mathematics, Panjab University, Chandigarh, India
e-mail: gkbakshi@yahoo.com, mraka@pu.ac.in, link2anu@gmail.com

**Abstract.** Let $q$ be an odd prime power and $m$ be a positive integer with $\gcd(m, q) = 1$. It is observed that the idempotent generators of irreducible cyclic codes of any length $m$ over the finite field $F_q$ can be effectively evaluated, once the same are known for irreducible cyclic codes of length $p^n$, where $p$ is any prime, even or odd and $n$ is a positive integer.

**Keywords.** Cyclic codes, cyclotomic cosets, idempotents.

## 1. Introduction

Let $F_q$ be the finite field with $q$ elements and $m$ be a positive integer coprime to $q$. A cyclic code $\mathcal{C}$ of length $m$ over $F_q$ is an ideal in the principal ideal ring $R_m = F_q[x]/\langle x^m - 1\rangle$. Every cyclic code $\mathcal{C}$ has a unique monic polynomial $g(x)$ of minimal degree, which generates the code and is also a divisor of $(x^m - 1)$, called the generator polynomial of $\mathcal{C}$. Besides the generator polynomial, every code $\mathcal{C}$ has a unique polynomial $e(x)$ that generates the code and also satisfies $(e(x))^2 = e(x)$, called the idempotent generator of $\mathcal{C}$. The two polynomials are related by $g(x) = \gcd\ (e(x), x^m - 1)$ computed in $F_q[x]$. An irreducible cyclic code of length $m$ over $F_q$ is a minimal ideal in $R_m$ and its idempotent generator is also called primitive idempotent. It is well known that every cyclic code is a direct sum of certain irreducible cyclic codes. Thus it is of interest to determine all the primitive idempotents in the ring $R_m$.

The generating idempotents of binary irreducible cyclic codes are easy to construct (see Van Lint [13, page 52]). The non-binary primitive idempotents have been obtained by various authors. Let $p$ be an odd prime and $q$ be an odd prime power, with $\gcd(p, q) = 1$. Berman [5] gave explicit expressions (without proof) of all the primitive idempotents in $R_{p^n}$, under the assumption that $q$ is a primitive root modulo $p^n$. Pruthi & Arora [9] verified Berman's result. Arora, Batra, Cohen & Pruthi [1] obtained all the primitive idempotents in $R_{p^n}$, when the multiplicative order of $q$ modulo $p^n$ is $\frac{\phi(p^n)}{2}$. Similar result was obtained in the ring $R_{2p^n}$ by Arora & Pruthi [2], when $q$ is a primitive root modulo $2p^n$. Bakshi & Raka [4] obtained all the primitive idempotents in $R_{2^n}$, $n \geq 3$, for any odd prime power $q$, $q \equiv 3, 5 \pmod 8$. In another paper, Bakshi & Raka [3] derived all the primitive idempotents in the ring $R_{p^n\ell}$, where $p$, $\ell$ are distinct odd primes, $q$ is a primitive root modulo $p^n$ and $\ell$ both, and $\gcd(\frac{\phi(p^n)}{2}, \frac{\phi(\ell)}{2}) = 1$.

*Research supported by N.B.H.M., India is gratefully acknowledged.

Let $q$ be an odd prime power, not necessarily a primitive root modulo $p^n$, where $p$ is an odd prime with $\gcd(p, q) = 1$. Let the multiplicative order of $q$ modulo $p$ be $f$ and $\gcd(\frac{p-1}{f}, q) = 1$. In [10,11], the authors gave efficient algorithms to compute explicit expressions of all the primitive idempotents in the ring $R_{p^n}$ and $R_{2p^n}$ for any positive integer $n$, provided $p$ does not divide $\frac{q^f-1}{p}$ if $n \geq 2$, thus generalizing the result of Berman [5] and of Arora & Pruthi [2], where $f = p-1$ and the result of Arora, Batra, Cohen & Pruthi [1], where $f = \frac{p-1}{2}$. In [12], the authors have also derived explicit expressions of all the primitive idempotents in the ring $R_{2^n}$, $n \geq 1$, where $q$ is any odd prime power, generalizing the result of Bakshi & Raka [4].

Now, a natural question arises:

**Can one determine the idempotent generators of irreducible cyclic codes of any length $m$, given the idempotent generators of those of prime power length?**

The answer is yes. In this paper, we prove this statement.The proof is similar to that of the construction of product codes, as given in MacWilliams & Sloane [8, Ch.18, §1–4].

Thus the problem of determination of the primitive idempotents in $R_m$, which was called non-constructive by Ding, Kohel and Ling [6], is almost completely solved. (The only case left is to determine the primitive idempotents in $R_{p^n}$, $n \geq 2$, when $\frac{q^f-1}{p}$ is divisible by $p$, where $p$ is an odd prime.)

## 2. Some preliminaries

Let $C$ be a cyclic code over $F_q$ with $g(x)$ as its generating polynomial. For $\alpha$, a fixed primitive $m$th root of unity in some extension field of $F_q$, the set $T = \{i \text{ (modulo } m) \mid g(\alpha^i) = 0\}$ is called the **defining set** of $C$.

For any integer $s$, the **$q$-cyclotomic coset** of $s$ modulo $m$ is the set

$$C_s^{(m)} = \{s, sq, sq^2, \ldots, sq^{m_s-1}\},$$

where $m_s$ is the least positive integer such that $sq^{m_s} \equiv s \pmod m$. The distinct $q$-cyclotomic cosets modulo $m$ partition the ring of integers $Z/mZ$ into disjoint sets. If $\alpha$ is a primitive $m$th root of unity in some extension field of $F_q$, then the minimal polynomial of $\alpha^s$ over $F_q$ is

$$M_{\alpha^s}(x) = \prod_{i \in C_s^{(m)}} (x - \alpha^i),$$

and $x^m - 1 = \prod_s M_{\alpha^s}(x)$ is the factorization of $x^m - 1$ into irreducible factors over $F_q$, where $s$ runs through a set of representatives of all the distinct $q$-cyclotomic cosets modulo $m$. Further corresponding to each cyclotomic coset $C_s^{(m)}$, the ideal generated by $(x^m - 1)/M_{\alpha^s}(x)$ is a minimal ideal in $R_m$, and is denoted by $\mathcal{M}_s^{(m)}$. Clearly the defining set of $\mathcal{M}_s^{(m)}$ is $Z/mZ - C_s^{(m)}$. The following results are well known. For reference, see MacWilliams & Sloane [8, Ch. 8] and Huffman & Pless [7].

**Lemma 1.** *If $e_s(x)$ denotes the idempotent generator of the irreducible cyclic code $\mathcal{M}_s^{(m)}$, then*

$$e_s(\alpha^j) = \begin{cases} 1 & \text{if } j \in C_s^{(m)}, \\ 0 & \text{if } j \notin C_s^{(m)}. \end{cases} \tag{1}$$

**Lemma 2.**

$$e_s(x) = \sum_{i=0}^{m-1} \epsilon_i^{(s)} x^i, \tag{2}$$

*where*

$$\epsilon_i^{(s)} = \frac{1}{m} \sum_{j \in C_s^{(m)}} \alpha^{-ij} \quad \textit{for} \quad 0 \le i \le m-1. \tag{3}$$

**Lemma 3.** *If $\mathcal{C}$ is a cyclic code of length $m$ generated by $g(x)$ and is of minimum distance $d$, then the code $\hat{\mathcal{C}}$ of length $mk$, generated by $g(x)(1 + x^m + x^{2m} + \cdots + x^{(k-1)m})$, is a repetition code of $\mathcal{C}$, repeated $k$ times, and its minimum distance is $dk$.*

A cyclic code is said to be **non-degenerate** if it is not a repetition of any code of smaller block length, otherwise the code is said to be degenerate.

**Lemma 4.** *An irreducible cyclic code $\mathcal{M}_s^{(m)}$ of length $m$ is nondegenerate if and only if* $\gcd(s, m) = 1$.

*Let $a$ be an integer such that* $\gcd(a, m) = 1$. *The function $\mu_a$, called a* **multiplier**, *denotes the coordinate permutation $i \mapsto ai \pmod m$, and acts on $R_m$ by*

$$\mu_a(f(x)) = f(x^a) (\text{mod } x^m - 1).$$

**Lemma 5.** *Let $\mathcal{C}$ be a cyclic code of length $m$ over $\mathrm{F}_q$ with the idempotent generator $e(x)$ and the defining set $T$. Let $a$ be an integer with* $\gcd(a, m) = 1$. *Then*

(i) *$\mu_a(e(x))$ is the idempotent generator of the equivalent cyclic code $\mu_a(\mathcal{C})$.*
(ii) *$a^{-1}T$ modulo $m$ is the defining set for $\mu_a(\mathcal{C})$, where $a^{-1}$ denotes the multiplicative inverse of $a$ modulo $m$.*

## 3. The main result

Firstly, in the following lemma, we see that it is enough to determine the idempotent generator of the non-degenerate irreducible cyclic code $\mathcal{M}_1^{(m)}$ for any positive integer $m$.

**Lemma 6.** *For any integer $s$ and a positive integer $m$,*

(i) *the idempotent generator $e_s(x)$ of the irreducible cyclic code $\mathcal{M}_s^{(m)}$ is given by*

$$\frac{1}{s_0} e_{s/s_0}(x)(1 + x^{\frac{m}{s_0}} + \cdots + x^{\frac{m}{s_0}(s_0-1)}),$$

*where $s_0 =$ $\gcd(s, m)$ and $e_{s/s_0}(x)$ is the idempotent generator of non-degenerate irreducible cyclic code $\mathcal{M}_{s/s_0}^{(m/s_0)}$.*

(ii) *if* $\gcd(s, m) = 1$, *then the idempotent generator $e_s(x)$ of non-degenerate irreducible cyclic code $\mathcal{M}_s^{(m)}$ is given by $e_s(x) = \mu_{s^{-1}}(e_1(x))$, where $e_1(x)$ is the idempotent generator of irreducible cyclic code $\mathcal{M}_1^{(m)}$.*

*Proof.* (i) Let $\beta = \alpha^{s_0}$, where $\alpha$ is a primitive $m$th root of unity in some extension field of $F_q$. Then it is easy to see that

$$\frac{x^m - 1}{M_{\alpha^s}(x)} = \frac{(x^{\frac{m}{s_0}} - 1)}{M_{\beta^{\frac{s}{s_0}}}(x)}(1 + x^{\frac{m}{s_0}} + \cdots + x^{\frac{m}{s_0}(s_0-1)}),$$

which implies, by Lemma 3, that the code $\mathcal{M}_s^{(m)}$ is a repetition code of the code $\mathcal{M}_{s/s_0}^{(m/s_0)}$, repeated $s_0$ times.

Let $e_s(x) = \frac{1}{s_0} e_{s/s_0}(x)(1 + x^{\frac{m}{s_0}} + \cdots + x^{\frac{m}{s_0}(s_0-1)})$. It is easy to see that the polynomial $e_s(x)$ satisfies $(e_s(x))^2 = e_s(x)$ in the ring $R_m$. Also we have $\gcd(x^{\frac{m}{s_0}} - 1, e_{s/s_0}(x)) = \frac{x^{\frac{m}{s_0}} - 1}{M_{\beta^{\frac{s}{s_0}}}(x)}$, which implies that $\gcd(e_s(x), x^m - 1) = \frac{x^m - 1}{M_{\alpha^s}(x)}$. Therefore $e_s(x)$ is the idempotent generator of the code $\mathcal{M}_s^{(m)}$.

(ii) If $\gcd(s, m) = 1$, we have $Z/mZ - C_s^{(m)} = \mu_s(Z/mZ - C_1^{(m)})$. Therefore by Lemma 5(ii), we get $\mathcal{M}_s^{(m)} = \mu_{s^{-1}}(\mathcal{M}_1^{(m)})$. Now the result follows from Lemma 5(i). □

Write $m = m_1 m_2$, where $m_1 \geq 1$, $m_2 \geq 1$ with $\gcd(m_1, m_2) = 1$. Therefore

$$Z/m_1Z \times Z/m_2Z \cong Z/mZ$$

as rings, and let the ring isomorphism be denoted by $\theta$.

We consider the following two cases separately.

I. $\gcd(O_{m_1}(q), O_{m_2}(q)) = 1$ and
II. $\gcd(O_{m_1}(q), O_{m_2}(q)) > 1$,

where $O_k(q)$ denotes the multiplicative order of $q$ modulo $k$.

**Theorem 1.** *Let* $m = m_1 m_2$, *where* $m_1 > 1$, $m_2 > 1$ *with* $\gcd(m_1, m_2) = 1$. *Suppose that* $\gcd(O_{m_1}(q), O_{m_2}(q)) = 1$. *Then the idempotent generator* $E(x)$ *of the irreducible cyclic code* $\mathcal{M}_1^{(m)}$ *of length* $m$ *over* $F_q$ *is given by*

$$E(x) = \chi(x^{m_2})\,\psi(x^{m_1}),$$

*where* $\chi(x)$ *and* $\psi(x)$ *are the idempotent generators of irreducible cyclic codes* $\mathcal{M}_1^{(m_1)}$ *and* $\mathcal{M}_1^{(m_2)}$ *over* $F_q$ *of lengths* $m_1$ *and* $m_2$ *respectively.*

*Proof.* It is clear that $C_1^{(m)} = \theta(C_1^{(m_1)} \times C_1^{(m_2)})$, as $\gcd(O_{m_1}(q), O_{m_2}(q)) = 1$.

Firstly, we claim that $\chi(x^{m_2})\psi(x^{m_1})$ is an idempotent in $R_m$.

As $\chi(x)$ is an idempotent in $R_{m_1} = F_q[x]/\langle x^{m_1} - 1\rangle$, so we have $(\chi(x))^2 \equiv \chi(x)(\text{mod } x^{m_1} - 1)$. This gives

$$(\chi(x^{m_2}))^2 \equiv \chi(x^{m_2})(\text{mod } x^m - 1). \tag{4}$$

Similarly, as $\psi(x)$ is an idempotent in $R_{m_2}$, so we have

$$(\psi(x^{m_1}))^2 \equiv \psi(x^{m_1})(\text{mod } x^m - 1). \tag{5}$$

From (4) and (5), we have

$$(\chi(x^{m_2})\,\psi(x^{m_1}))^2 \equiv \chi(x^{m_2})\,\psi(x^{m_1})(\bmod x^m - 1).$$

That is, $\chi(x^{m_2})\,\psi(x^{m_1})$ is an idempotent in $R_m$.

If $\alpha$ is a primitive $m$th root of unity in some extension field of $F_q$, then $\beta = \alpha^{m_2}$ is a primitive $m_1$th root of unity and $\gamma = \alpha^{m_1}$ is a primitive $m_2$th root of unity.

By Lemma 1, we see that $\chi(\alpha^{jm_2})\,\psi(\alpha^{jm_1}) = \chi(\beta^j)\psi(\gamma^j) = 0$ or $1$ for $j = 0, 1, \cdots, m-1$, as $\chi(x)$ and $\psi(x)$ are primitive idempotents in the rings $R_{m_1}$ and $R_{m_2}$ respectively.

Now $\chi(\alpha^{jm_2})\,\psi(\alpha^{jm_1}) = 1$ if and only if $\chi(\beta^j) = \psi(\gamma^j) = 1$ if and only if $j \in C_1^{(m_1)}$ and $j \in C_1^{(m_2)}$ if and only if $j \in \theta(C_1^{(m_1)} \times C_1^{(m_2)}) = C_1^{(m)}$.

This proves the theorem. □

**Theorem 2.** *Let $m = m_1 m_2$, where $m_1 > 1$, $m_2 > 1$ with $\gcd(m_1, m_2) = 1$. Suppose that $\gcd(O_{m_1}(q), O_{m_2}(q)) = s > 1$. Then the generating idempotent $E(x)$ of the irreducible cyclic code $\mathcal{M}_1^{(m)}$ of length $m$ over $F_q$ is given by*

$$E(x) = T_s(\hat{\chi}(x^{m_2})\hat{\psi}(x^{m_1})),$$

*where $\hat{\chi}(x)$ and $\hat{\psi}(x)$ are the generating idempotents of irreducible cyclic codes $\hat{\mathcal{M}}_1^{(m_1)}$ and $\hat{\mathcal{M}}_1^{(m_2)}$ over $F_{q^s}$ of lengths $m_1$ and $m_2$ respectively, and $T_s : F_{q^s}[x] \mapsto F_q[x]$ is the trace function, defined by*

$$T_s\left(\sum_{i=0}^{m-1} a_i x^i\right) = \sum_{i=0}^{m-1} T_s(a_i)x^i = \sum_{i=0}^{m-1}\sum_{j=0}^{s-1} a_i^{q^j} x^i \quad \text{for} \quad \sum_{i=0}^{m-1} a_i x^i \in F_{q^s}[x].$$

*Proof.* Write

$$C_1^{(m)} = \hat{C}_1^{(m)} \cup \hat{C}_q^{(m)} \cup \cdots \cup \hat{C}_{q^{s-1}}^{(m)},$$

where $\hat{C}_{q^u}^{(m)} = \{q^u, q^{u+s}, \ldots, q^{u+s(\frac{k}{s}-1)}\}$ for $0 \le u \le s-1$, is the $q^s$-cyclotomic coset of $q^u$ modulo $m$ with $k = O_m(q)$.

By Lemma 2, we have $E(x) = \sum_{i=0}^{m-1} \epsilon_i x^i$, where

$$\epsilon_i = \frac{1}{m}\sum_{j\in C_1^{(m)}} \alpha^{-ij} = \frac{1}{m}\sum_{u=0}^{s-1}\sum_{j\in \hat{C}_{q^u}^{(m)}} \alpha^{-ij} = \frac{1}{m}\sum_{u=0}^{s-1}\left(\sum_{j\in \hat{C}_1^{(m)}} \alpha^{-ijq^u}\right)$$

$$= \frac{1}{m}\sum_{u=0}^{s-1}\left(\sum_{j\in \hat{C}_1^{(m)}} \alpha^{-ij}\right)^{q^u} = T_s\left(\frac{1}{m}\sum_{j\in \hat{C}_1^{(m)}} \alpha^{-ij}\right). \tag{6}$$

If $\hat{E}(x)$ is the generating idempotent of the irreducible cyclic code $\hat{\mathcal{M}}_1^{(m)}$ of length $m$ over $F_{q^s}$ whose defining set is $Z/mZ - \hat{C}_1^{(m)}$, then again by Lemma 2, we have

$$\hat{E}(x) = \sum_{i=0}^{m-1} \hat{\epsilon}_i x^i, \quad \text{where } \hat{\epsilon}_i = \frac{1}{m}\sum_{j\in \hat{C}_1} \alpha^{-ij}. \tag{7}$$

From (6) and (7), we have $\epsilon_i = T_s(\hat{\epsilon}_i)$. This gives

$$E(x) = \sum_{i=0}^{m-1} T_s(\hat{\epsilon}_i)x^i = T_s\left(\sum_{i=0}^{m-1} \hat{\epsilon}_i x^i\right) = T_s(\hat{E}(x)).$$

Since $\gcd(O_{m_1}(q^s), O_{m_2}(q^s)) = \frac{1}{s}\gcd(O_{m_1}(q), O_{m_2}(q)) = 1$, so by Theorem 1, we have $\hat{E}(x) = \hat{\chi}(x^{m_2})\hat{\psi}(x^{m_1})$, which completes the proof of the theorem. □

## References


[1] S. K. Arora, S. Batra, S. D. Cohen and M. Pruthi, The primitive idempotents of a cyclic group algebra, *Southeast Asian Bulletin of Mathematics* **26** no. 4 (2002) 549–557

[2] S. K. Arora and M. Pruthi, Minimal cyclic codes of length $2p^n$, *Finite Fields and their Appl.* **5** no. 2 (1999) 177–187

[3] G. K. Bakshi and M. Raka, Minimal cyclic codes of length $p^n q$, *Finite Fields and their Appl.* **9** no. 4 (2003) 432–448

[4] G. K. Bakshi and M. Raka, Minimal cyclic codes of length $2^m$, *Ranchi University Math. Journal* **33** (2002) 1–18

[5] D. Berman, Semisimple cyclic and abelian code. II. Cybernetics **3** (3) (1967) 17–23

[6] C. Ding, D. R. Kohel and S. Ling, Split group codes, *IEEE Trans. Inform. Theory* 46, no. 2 (2000) 485–495

[7] W. C. Huffman and V. Pless, Fundamentals of Error-Correcting Codes, Cambridge University Press (2003)

[8] F. J. MacWilliams and N. J. A. Sloane, Theory of Error-Correcting Codes, North-Holland, Amsterdam (1977)

[9] M. Pruthi and S. K. Arora, Minimal codes of prime-power length, *Finite Fields and their Appl.* **3** no. 2 (1997) 99–113

[10] A. Sharma, G. K. Bakshi, V. C. Dumir and M. Raka, Cyclotomic numbers and primitive idempotents in the ring $GF(q)[x]/\langle x^{p^n} - 1\rangle$, *Finite fields and their Appl.* **10** no. 4 (2004) 653–673

[11] A. Sharma, G. K. Bakshi, V. C. Dumir and M. Raka, Irreducible cyclic codes of length $2^n$, *ARS Combinatoria* **86** (2008) 133–146

[12] A. Sharma, G. K. Bakshi and M. Raka, Irreducible cyclic codes of length $2p^n$, *ARS Combinatoria* **83** (2007) 267–278

[13] J. H. Van Lint, Coding Theory, Springer-Verlag, Berlin. Heidelberg. New York, **201** (1971)

Proc. Int. Conf. – Number Theory and Discrete Geometry No. 6, 2008, pp. 19–34.

# On the Partial Fraction Expansion for the Cotangent-Like Function

R. Balasubramanian[1], L. P. Ding[2], S. Kanemitsu[3] and Y. Tanigawa[4]

[1]The Institute of Mathematical Sciences, Chennai 600 133, India
e-mail: director@imsc.res.in
[2]School of Science, Xi'an University of Architecture & Technology, Xi'an 710055, P.R. China
e-mail: lpding@123.com
[3]Kinki University School of Humanity-Oriented Science and Engineering, Iizuka, Fukuoka, 820-8555, Japan
e-mail: kanemitu@fuk.kindai.ac.jp
[4]Graduate School of Mathematics, Nagoya University, Nagoya, 464-8602, Japan
e-mail: tanigawa@math.nagoya-u.ac.jp

## 1. Introduction and statement of results

The partial fraction expansion referred to in the title is the following

$$2\pi i\left(\frac{e^{2\pi i\alpha mz}}{e^{2\pi imz}-1}+\frac{\delta(\alpha)}{2}\right)-\frac{1}{mz}=\sum_{\substack{n=-\infty\\ n\neq 0}}^{\infty}\frac{e^{-2\pi in\alpha}}{n+mz} \tag{1.1}$$

valid for all $z\in\mathbb{C}$, except for those which give rise to the singularities of the members, so that (1.1) is valid for $z\in\mathbb{C}\setminus\mathbb{Q}$, where $\delta(\alpha)=1$ for $\alpha=1$ and $\delta(\alpha)=0$ otherwise. The sum is taken in the Cauchy mean value sense. (1.1) may be conveniently expressed as

$$2\pi i\left(\frac{e^{2\pi i\alpha mz}}{e^{2\pi imz}-1}+\frac{\delta(\alpha)}{2}\right)=\sum_{n=-\infty}^{\infty}\frac{e^{2\pi in\alpha}}{mz-n} \tag{1.2}$$

which reduces in the case $m=1$ and $\alpha=1$ to the partial fraction expansion for the cotangent function

$$\pi\cot\pi z=\sum_{n=-\infty}^{\infty}\frac{1}{n+z},\qquad z\in\mathbb{C}\backslash\mathbb{Z}, \tag{1.3}$$

on noting that

$$2\pi i\left(\frac{e^{2\pi iz}}{e^{2\pi iz}-1}-\frac{1}{2}\right)=\pi\cot\pi z. \tag{1.4}$$

(1.3) is equivalent to that for the hyperbolic cotangent function under the transition ($i\tau = z$; $\tau \in \mathbb{H} = \{\tau \in \mathbb{C} \mid \mathrm{Im}\tau > 0\} \longleftrightarrow z \in \{z \in \mathbb{C} \mid \mathrm{Re}z > 0\}$)

$$\pi \coth \pi z = \frac{2\pi}{e^{2\pi z} - 1} + \pi = \frac{1}{z} + 2z \sum_{n=1}^{\infty} \frac{1}{n^2 + z^2}, \quad \mathrm{Re}z \geq 0. \tag{1.5}$$

(1.5) is used to deduce the functional equation ([26])

$$\pi^{-\frac{s}{2}} \Gamma\left(\frac{s}{2}\right) \zeta(s) = \pi^{-\frac{1-s}{2}} \Gamma\left(\frac{1-s}{2}\right) \zeta(1-s) \tag{1.6}$$

for the Riemann zeta-function

$$\zeta(s) = \sum_{n=1}^{\infty} \frac{1}{n^s}, \quad \sigma = \mathrm{Re}s > 1, \tag{1.7}$$

which is continued meromorphically over the whole plane through (1.6).

On the other hand, (1.5) is deduced from (1.6) in [19] and it follows that (1.3), (1.5) and (1.6) are equivalent as noticed by [17] and [28].

The counterpart of (1.2) reads

$$2\pi i \left( \frac{e^{2\pi\alpha m z}}{e^{2\pi m z} - 1} + \frac{\delta(\alpha)}{2} \right) = - \sum_{n=-\infty}^{\infty} \frac{e^{2\pi i n \alpha}}{n + imz} \tag{1.8}$$

valid for all $z$ except for the zeros of the denominators.

Although (1.8) was used at least three times in the literature [12], [3], [27] in different contexts and without noticing its source, we are going to streamline the situation and deduce everything from the functional equation (1.6) only.

What we use is a seemingly more general functional equation for the Hurwitz zeta-function

$$\zeta(s, x) = \sum_{n=0}^{\infty} \frac{1}{(n+x)^s}, \quad \sigma = \mathrm{Re}s > 1, \ x > 0. \tag{1.9}$$

Note that $\zeta(s, 1) = \zeta(s)$.

The functional equation for $\zeta(s, x)$ (or sometimes referred to as Hurwitz's or Lerch's relation) reads

$$\zeta(1-s, x) = \frac{\Gamma(s)}{(2\pi)^s} (e^{-\frac{\pi i}{2} s} l_s(x) + e^{\frac{\pi i}{2} s} l_s(1-x)), \quad 0 < x < 1 \tag{1.10}$$

with $l_s(x)$ designating the polylogarithm function of the complex exponential argument

$$l_s(x) = \sum_{n=1}^{\infty} \frac{e^{2\pi i n x}}{n^s}, \quad \sigma > 1, \tag{1.11}$$

which is defined also for all $s \in \mathbb{C}$ and $x \in \mathbb{H}$, the upper half-plane.

We remark that this polylogarithm function or rather its generalization, the Lerch (or the Hurwitz-Lerch) zeta-function is a missing link that connects $\mathbb{H}$ and the right half-plane $\operatorname{Re} z > 0$ (which inherits the properties of functions considered on the positive real axis).

The Lerch zeta function $\phi(\xi, s, x)$ is defined by

$$\phi(\xi, s, x) = \sum_{n=0}^{\infty} \frac{e^{2\pi i n\xi}}{(n+x)^s}, \quad \sigma > 1 \tag{1.12}$$

(cf. [9], [20], [15]).

In [15, Example 2], (1.1) is deduced from the functional equation for $\phi(\xi, s, x)$ and its Taylor expansion in $\xi$, on appealing to

$$\zeta(-m, x) = -\frac{B_{m+1}(x)}{m+1}, \qquad m = 0, 1, \ldots, \tag{1.13}$$

where $B_{m+1}(x)$ indicates the $m+1$ st Bernoulli polynomials. It is an instructive exercise to deduce the functional equation for $\phi$ from (1.1) as has been done in [16]. Invoking Example 6 in [18] to the effect that (1.6) and (1.10) are equivalent, we may now have recourse to (1.8) as a substitute for (1.6). For the sake of completeness we shall derive the functional equation (1.10) from (1.6) in §3.

The "Fourier expansion" (1.8) has an interesting history as referred to above. We shall now describe its use in the context of the celebrated transformation formula for the Dedekind eta-function $\eta(\tau)$

$$\eta(\tau) = q^{\frac{1}{24}} \prod_{n=1}^{\infty} (1 - q^n), \quad q = e^{2\pi i n\tau}, \tau \in \mathbb{H} \tag{1.14}$$

under the modular substitution. It was S. Iseki [12] who first applied (1.8) in this context. To describe his results, we introduce the notation. Let

$$\lambda_p(x) = \sum_{m=0}^{\infty} m^{-p} e^{-2\pi m x}, \quad \operatorname{Re} x > 0 \tag{1.15}$$

and introduce the Lambert series

$$\Lambda_p(\alpha, \beta, z) = \sum_{l=0}^{\infty} \left\{ \lambda_p((l+\alpha)z - i\beta) + \lambda_p((l+1-\alpha)z + i\beta) \right\} \tag{1.16}$$

for an odd integer $p \geq 1$.

Then Iseki's transformation formula reads

$$\Lambda_p(\alpha, \beta, z) = (iz)^{p-1} \Lambda_p(1-\beta, \alpha, z^{-1}) + R_p(\alpha, \beta, z), \tag{1.17}$$

where

$$R_p(\alpha, \beta, z) = -\frac{(2\pi z)^p}{(p+1)!} \sum_{n=0}^{p+1} \binom{p+1}{n} (iz)^{-n} \overline{B}_{p+1-n}(\alpha) \overline{B}_n(\beta), \tag{1.18}$$

where $\overline{B}_\nu(x)$ is the periodic Bernoulli polynomial defined by $B_\nu(x-[x])$, with $[x]$ indicating the integral part of $x$. For its Fourier expansion, see (1.26) below.

As is pointed out by Iseki (and by Mikolás [22]), Formula (1.17) is a refinement (and correction) of Apostol's transformation formula for ($x>0,\ p>1$ odd integer)

$$F_p(x)=\sum_{n=1}^{\infty}\lambda_p(nx)=\sum_{n=1}^{\infty}\sum_{m=1}^{\infty}\frac{1}{m^p}e^{-2\pi xmn}: \tag{1.19}$$

$$F_p\left(\frac{z-ih}{k}\right)+\frac{1}{2}\zeta(p)(1-(-1)^{\frac{1}{2}(p-1)}z^{p-1})$$
$$+\frac{(2\pi z)^p}{2(p+1)!}\sum_{\nu=0}^{p+1}\binom{p+1}{\nu}(iz)^{-\nu}\sum_{\mu=1}^{k}\overline{B}_{p+1-\nu}\left(\frac{\mu}{k}\right)\overline{B}_\nu\left(\frac{h\mu}{k}\right)$$
$$=(-1)^{\frac{p-1}{2}}z^{p-1}F_p\left(\frac{1/z-iH}{k}\right) \tag{1.20}$$

for $hH\equiv -1 \pmod k$, in which the term containing $\zeta(p)$ was missing in Apostol [1, Theorem 2], and which is valid in the limiting case $p\longrightarrow 1$ as we shall see below. The case $p=1$ was considered by Rademacher [24], giving rise to the transformation formula for $\eta(\tau)$

$$\eta\left(-\frac{1}{\tau}\right)=\left(\frac{\tau}{i}\right)^{\frac{1}{2}}\eta(\tau).$$

We note that (1.20) with $k=1, h=H=0$ yields the famous Ramanujan's formula for the values of $\zeta(s)$ at odd integral arguments $p>1$ as noted by Apostol, Grosswald [11] and Berndt [5].

Indeed, for $p>1$ we have

$$F_p(x)=\sum_{m=1}^{\infty}\frac{e^{-2\pi xm}}{m^p(1-e^{-2\pi xm})}=\sum_{m=1}^{\infty}\frac{m^{-p}}{e^{2\pi xm}-1}\left(=\sum_{k=1}^{\infty}\sigma_{-p}(k)e^{-2\pi xk}\right),$$

so that (1.20) reads (on writing $p=2n+1$)

$$(\pi x)^{-n}\left\{\frac{1}{2}\zeta(2n+1)+\sum_{m=1}^{\infty}\frac{m^{-2n-1}}{e^{2\pi xm}-1}\right\}$$
$$=\left(-\frac{x}{\pi}\right)^{-n}\left\{\frac{1}{2}\zeta(2n+1)+\sum_{m=1}^{\infty}\frac{m^{-2n-1}}{e^{2\pi\frac{m}{x}}-1}\right\}$$
$$-2^{2n}\sum_{j=0}^{n-1}(-1)^j\frac{B_{2j}}{(2j)!}\frac{B_{2n+2-2j}}{(2n+2-2j)!}(\pi x)^{n+1-j}\left(\frac{x}{\pi}\right)^j \tag{1.21}$$

(cf. e.g. [5] and [14]).

Then Iseki [13] gave a "simpler" proof of (1.17) in the case $p=1$ by Rademacher's [24] Mellin transform method, i.e. the modular relation. Apostol [2] then gave the proof of (1.17) for $p > 1$ odd by considering the generating zeta-function

$$Z(\alpha, \beta, s) = \zeta(s, \alpha)\ell_{p+s}(\beta) + \zeta(s, 1-\alpha)\ell_{p+s}(1-\beta),$$

which satisfies the functional equation of Hecke type: putting $F_p(\alpha, \beta, s) = \frac{\Gamma(s)}{(2\pi)^s} Z(\alpha, \beta, s)$, then

$$F_p(\alpha, \beta, 1-p-s) = i^{p-1} F_p(1-\beta, \alpha, s).$$

Then Basu [3], overlooking Iseki's first paper [12], states "Both Iseki's original derivation of (1.17), and a shorter proof given by Apostol [2] use complex integration and residue calculus," which is to be corrected according to the preceding lines. Employing Basu's notation, Iseki's formula (1.17) may be expressed in a more lucid and natural way.

We introduce the (lattice) Lambert series

$$G_p(\alpha, \beta, z) = \sum_{m=-\infty}^{\infty}{}' \sum_{n=-\infty}^{\infty}{}' \frac{e^{2\pi i m\beta} e^{2\pi i m\alpha}}{m^p(n - izm)}. \tag{1.22}$$

Then (1.17) reads

$$G_p(\alpha, \beta, z) = (iz)^{p-1} G_p\left(1-\beta, \alpha, \frac{1}{z}\right) + \sum_{r=1}^{p} (iz)^{p-r} E_{p-r+1}(\alpha) E_r(\beta), \tag{1.23}$$

where

$$E_\nu(x) = -\frac{(2\pi i)^\nu}{\nu!} B_\nu(x) \tag{1.24}$$

and where $B_\nu(x)$ signifies the $\nu$-th Bernonlli polynomial, since

$$\Lambda_p(\alpha, \beta, z) = \frac{1}{2\pi i} G_p(\alpha, \beta, z) + \frac{1}{2\pi i z} E_{p+1}(\beta). \tag{1.25}$$

Iseki's argument depends on (1.2), the inversion of the order of summation and the Fourier expansion for the periodic Bernoulli polynomial (cf. (1.24))

$$\overline{B}_\nu(x) = -\frac{\nu!}{(2\pi i)^\nu} \sum_{\substack{n=-\infty \\ n\neq 0}}^{\infty} \frac{e^{2\pi i n x}}{n^\nu}. \tag{1.26}$$

After Iseki [12], it was Sczech [25] who applied (1.3) and the identity (from the theory of modular forms)

$$-4\pi i \frac{d}{dz} \log \eta(z) = \sum_{m} \sum_{\substack{n \\ (m,n)\neq(0,0)}} (mz+n)^{-2}, \tag{1.27}$$

the left-hand side being equal to

$$2\zeta\left(\frac{1}{2}\right) = \frac{\pi^2}{3} + 2\pi^2 \sum_{m=1}^{\infty} \frac{1}{\sin^2 \pi m z}$$

(where $\zeta(z)$ is the Weierstrass zeta-function, cf. Elstrodt [8]). He made the inversion of the order of summation, calling it the Eisenstein summation. He refers to (1.8) as an analytic continuation of the Fourier series. For the history of the transformation formula for $\eta(\tau)$, we refer to [14].

Finally, we mention the paper of Goldstein and de la Torre [10] which gives the general transformation formula for $\eta(\tau)$ for a general $\sigma \in SL_2(\mathbb{Z})$ by streamlining the argument of Rademacher [24] in the spirit of Weil [29]. The relevant zeta-function is

$$Z(s) = Z(s, \varepsilon, p, q, c) = \zeta_\varepsilon(s, p, c)\zeta_\varepsilon(s+1, q, c),$$

where

$$\zeta_\varepsilon(s, h, k) = \sum_{\substack{n=-\infty \\ n\equiv h (\operatorname{mod} k)}}^{\infty}{}' \frac{\nu_\varepsilon(n)}{|n|^s} \tag{1.28}$$

and where $\nu_\varepsilon(n)$ in the sign character defined by

$$\nu_\varepsilon(n) = \left(\frac{n}{|n|}\right)^\varepsilon, \quad \varepsilon = 0 \text{ or } 1.$$

The zeta-function $\zeta_\varepsilon$ satisfies the functional equation

$$\left(\frac{\pi}{k}\right)^{-\frac{s}{2}} \Gamma\left(\frac{s+\varepsilon}{2}\right) \zeta_\varepsilon(s, h, k)$$
$$= (-i)^\varepsilon k^{-\frac{1}{2}} \left(\frac{\pi}{k}\right) \Gamma\left(\frac{1-s+\varepsilon}{2}\right) \sum_{\lambda \operatorname{mod} k} e^{2\pi i \frac{h}{k}\lambda} \zeta_\varepsilon(1-s, \lambda, k). \tag{1.29}$$

The essential point here, which also prevails all the researches about the Lambert series (1.16), is that the zeta-function satisfies the functional equation of ramified type (with a single gamma factor).

It would be a challenging problem to consider what one will get when one considers the case where $p$ is even. Then necessarily we have to appeal to the case of functional equation with two gamma factors (cf. [18, Chap. 3]).

We are in a position to state

**Theorem 1.** *Iseki's transformation formula (1.17), and a fortiori, the eta transformation formula is a consequence of the functional equation (1.6) for the Riemann zeta-function.*

From what precedes, we have

$$(1.6) \Longleftrightarrow (1.10) \Longleftrightarrow (1.8), \quad (1.8) + (1.26) \Longrightarrow (1.17).$$

Hence it is enough to deduce (1.26) from (1.10), which depends on how we define the Bernoulli polynomial. If we define $(0 < x < 1)$

$$B_s(x) = -s\zeta(1-s, x) \tag{1.30}$$

as in Milnor [23] or Campbell [6], we can prove that $B_n(x)$ is a polynomial of degree $n$ that satisfies the same conditions as the Bernoulli polynomial. Then for $s = 1$, (1.10) gives

$$\begin{aligned}\overline{B}_1(x) &= -\frac{1}{2\pi}\sum_{n=1}^{\infty}\frac{e^{2\pi inx}}{n}e^{-\frac{\pi i}{2}} + \sum_{n=1}^{\infty}\frac{e^{2\pi in(1-x)}}{n}e^{\frac{\pi i}{2}} \\ &= -\frac{1}{2\pi i}\left(\sum_{n=1}^{\infty}\frac{e^{2\pi inx}}{n} + \sum_{n=-\infty}^{-1}\frac{e^{2\pi inx}}{n}\right)\end{aligned}$$

which is the special case of (1.26) (convergence follows from Dirichlet's test) and the general case follows on integration.

## 2. Eisenstein's formula

Eisenstein's formula appears wherever there is need to express $\overline{B}_1(x)$ in terms of the cotangent function:

$$\overline{B}_1\left(\frac{a}{q}\right) = -\frac{1}{2q}\sum_{k=1}^{q-1}\sin\frac{2ka}{q}\pi\cot\frac{k}{q}\pi. \tag{2.1}$$

This and its generalization due to K. Wang has recently been structurally elucidated by Li–Hashimoto–Kanemitsu [21] as a relation between two bases of the vector space of periodic function with period $q$.

One of the theorems (Theorem E) of Wang escapes the elucidation, and we shall derive it from (1.2) and (1.4).

**Theorem 2 (Wang [27]).** *Let $q > 2$ be a fixed integer and let $a$ be an integer $\not\equiv 0 \pmod q$. Then*

$$\frac{1}{2\pi i}\sum_{k=1}^{q-1}e^{-2\pi i\frac{ak}{q}}\pi\cot\pi\left(\frac{x}{\pi}+\frac{k}{q}\right) = q\frac{e^{2iax}}{e^{2iqx}-1} - \frac{e^{2ix}}{e^{2ix}-1} + \frac{1}{2}. \tag{2.2}$$

*Proof.* By (1.3), the LHS, say $S$, is

$$\begin{aligned}\frac{1}{2\pi i}\sum_{k=1}^{q-1}\sum_{n=-\infty}^{\infty}\frac{e^{2\pi i\frac{a}{q}k}}{n+\frac{x}{\pi}+\frac{k}{q}} &= \frac{q}{2\pi i}\sum_{k=1}^{q-1}\sum_{n=-\infty}^{\infty}\frac{e^{2\pi i\frac{a}{q}(k+nq)}}{q\frac{x}{\pi}+k+nq} \\ &= \frac{q}{2\pi i}\sum_{k=1}^{q-1}\sum_{\substack{m=-\infty \\ m\equiv k \pmod q}}^{\infty}\frac{e^{2\pi i\frac{a}{q}m}}{q\frac{x}{\pi}+m}.\end{aligned}$$

Supplementing the missing term with $k = 0$, i.e.

$$\sum_{n=-\infty}^{\infty}\frac{1}{q\frac{x}{\pi}+nq},$$

which is $\frac{\pi}{q}\cot\pi\frac{x}{\pi}$, we conclude that

$$S = \frac{q}{2\pi i}\sum_{k=0}^{q}\sum_{\substack{m=-\infty\\ m\equiv k(\mathrm{mod}\, q)}}^{\infty}\frac{e^{2\pi i\frac{a}{q}m}}{q\frac{x}{\pi}+m} - \frac{1}{2i}\cot x.$$

Since the first double sum is the decomposition of integers into residue classes, we have

$$\begin{aligned} S &= \frac{q}{2\pi i}\sum_{n=-\infty}^{\infty}\frac{e^{2\pi i\frac{a}{q}n}}{q\frac{x}{\pi}+n} - \frac{1}{2i}\cot x \\ &= \frac{q}{2\pi i}2\pi i\frac{e^{2\pi i\frac{a}{q}q\frac{x}{\pi}}}{e^{2\pi i q\frac{x}{q}}-1} - \frac{e^{2ix}}{e^{2ix}-1}+\frac{1}{2} \end{aligned}$$

by (1.2) and (1.4), where (2, 2) follows.

*Remark.* Wang's theorem C, which is a generalization of (2.1), follows from (2.2) on recalling

$$\frac{ze^{xz}}{e^z-1} = \sum_{n=0}^{\infty}\frac{B_n(x)}{n!}z^n, \quad |z|<2\pi$$

and in particular

$$\frac{z}{e^z-1} = \sum_{n=0}^{\infty}\frac{B_n}{n!}z^n, \quad B_n = B_n(0).$$

## 3. The functional equation the Hurwitz zeta-function

The functional equation (1.10) the Hurwitz zeta-function is usually proved by generalizing Riemann's first proof for that for the Riemann zeta-function ([26, 2.17]) or by the partial fraction expansion for cot-like function ([16]). In this section we shall directly deduce (1.10) from (1.6).

**Theorem 3.** *The functional equation (1.6) for the Riemann zeta-function implies the functional equation (1.10) for the Hurwitz zeta-function, so that they are equivalent.*

First we shall recall the general modular relation from [17]. Let $\{\lambda_k\}_{k=1}^{\infty}$, $\{\mu_k\}_{k=1}^{\infty}$ be increasing sequences tending to $\infty$ and let $\{\alpha_k\}_{k=1}^{\infty}$, $\{\beta_k\}_{k=1}^{\infty}$ be complex sequences. We form the Dirichlet series $\phi(s)$ and $\psi(s)$ from this data, and suppose they are absolutely convergent with abscissas of absolute convergence $\sigma_\phi$ and $\sigma_\psi$:

$$\phi(s) = \sum_{k=1}^{\infty}\frac{\alpha_k}{\lambda_k^s}, \quad \sigma>\sigma_\phi$$

$$\psi(s) = \sum_{k=1}^{\infty}\frac{\beta_k}{\mu_k^s}, \quad \sigma>\sigma_\psi.$$

We say that $\phi(s)$ and $\psi(s)$ satisfy the functional equation with multiple gamma factors and write

$$\frac{\Gamma(\{d_j + D_j s\}_{j=1}^{M})}{\Gamma(\{c_j + C_j s\}_{j=1}^{P})}\phi(s) = \frac{\Gamma(\{e_j + E_j(r-s)\}_{j=1}^{\tilde{N}})}{\Gamma(\{f_j + F_j(r-s)\}_{j=1}^{\tilde{Q}})}\psi(r-s)$$

where $r$ is a real number, coefficients $C_j$ etc. in capital letters are all positive, coefficients in small letters may be complex numbers and $\Gamma(\{k_j\}_{j=1}^{K})$ is an abbreviation of $\prod_{j=1}^{K}\Gamma(k_j)$, if there exist a meromorphic function $\tilde{\chi}(s)$ admitting the representations:

$$\tilde{\chi}(s) = \frac{\Gamma(\{d_j + D_j s\}_{j=1}^{M})}{\Gamma(\{c_j + C_j s\}_{j=1}^{P})}\phi(s), \quad \text{for} \quad \sigma > \sigma_\phi$$

and

$$\tilde{\chi}(s) = \frac{\Gamma(\{e_j + E_j(r-s)\}_{j=1}^{\tilde{N}})}{\Gamma(\{f_j + F_j(r-s)\}_{j=1}^{\tilde{Q}})}\psi(r-s), \quad \text{for} \quad \sigma < r - \sigma_\psi,$$

with a finite number of poles with a certain growth condition.

We rewrite the above functional equation in the following form:

$$\chi(s) = \phi(s) = G(s)\psi(r-s), \tag{3.1}$$

with a finite number of poles $s_k$, $1 \le k \le L$, all contained in a compact set $S$, and subject to the growth condition

$$\lim_{|v|\to\infty} \Delta(u + i\, v)\chi(u + i\, v) = 0$$

uniformly in $u$, $u_1 \le u \le u_2$, for any $u_1 < u_2$, where

$$G(s) = \frac{\Gamma(\{e_j + E_j(r-s)\}_{j=1}^{\tilde{N}})}{\Gamma(\{f_j + F_j(r-s)\}_{j=1}^{\tilde{Q}})}\frac{\Gamma(\{c_j + C_j s\}_{j=1}^{P})}{\Gamma(\{d_j + D_j s\}_{j=1}^{M})}.$$

Let $\Delta(s)$ be the "processing gamma factor" defined by

$$\Delta(s) = \frac{\Gamma(\{b_j + B_j s\}_{j=1}^{m}, \{a_j - A_j s\}_{j=1}^{n})}{\Gamma(\{a_j + A_j s\}_{j=n+1}^{p}, \{b_j - B_j s\}_{j=m+1}^{q})}.$$

Associated with $\Delta(s)$, we define the Fox $H$-function

$$H_{p,q}^{m,n}\left(z \,\middle|\, \begin{matrix} \{(1-a_j, A_j)\}_{j=1}^{n}, \{(a_j, A_j)\}_{j=n+1}^{p} \\ \{(b_j, B_j)\}_{j=1}^{m}, \{(1-b_j, B_j)\}_{j=n+1}^{q} \end{matrix}\right)$$
$$= \frac{1}{2\pi i}\int_{L_1} \frac{\Gamma(\{(b_j + B_j s)\}_{j=1}^{m}, \{(a_j - A_j s)\}_{j=1}^{n})}{\Gamma(\{(a_j + A_j s)\}_{j=n+1}^{p}, \{(b_j - B_j s)\}_{j=m+1}^{q})} z^{-s}\, ds,$$

where $z \notin (-\infty, 0]$ and $L_1$, sometimes denoted by $(\gamma_1)$, means the deformed Bromwich contour $L_1 : s = \gamma_1 + i\,t, \quad -\infty < t < \infty$ with a possible indentation in the finite part as described below so that it separates all the poles of the gamma factors: We take two contours $L_1 = (\gamma_1)$ and $L_2 = (\gamma_2)$ $(\gamma_2 < \gamma_1)$ and suppose that they squeeze the above-mentioned compact set $S$, with the boundary $C$, and that all the poles of

$$\frac{\Gamma(\{b_j + B_j s\}_{j=1}^{m})}{\Gamma(\{a_j + A_j s\}_{j=n+1}^{p}, \{b_j - B_j s\}_{j=m+1}^{q})}$$

lie to the left of $L_2$ and those of

$$\frac{\Gamma(\{a_j - A_j s\}_{j=1}^{n})}{\Gamma(\{a_j + A_j s\}_{j=n+1}^{p}, \{b_j - B_j s\}_{j=m+1}^{q})}$$

lie to the right of $L_1$, so that *a fortiori*, $L_1$ and $L_2$ separate the poles of the gamma factors. When we speak of taking $L_1$ to the right of $\sigma_\phi$ $(\gamma_1 > \sigma_\phi)$ and $L_2$ to the left of $r - \sigma_\psi$ $(\gamma_2 < r - \sigma_\psi)$, we mean the above separation conditions are satisfied and in the latter case also that all the poles of

$$\frac{\Gamma(\{b_j + B_j s\}_{j=1}^{m}, \{c_j + C_j s\}_{j=1}^{P})}{\Gamma(\{a_j + A_j s\}_{j=n+1}^{p}, \{b_j - B_j s\}_{j=m+1}^{q}, \{f_j + F_j r - F_j s\}_{j=1}^{\tilde{Q}}, \{d_j + D_j\}_{j=1}^{M})}$$

lie to the left of $L_2$ and those of

$$\frac{\Gamma(\{a_j - A_j s\}_{j=1}^{n}, \{e_j + E_j r - E_j s\}_{j=1}^{\tilde{N}})}{\Gamma(\{a_j + A_j s\}_{j=n+1}^{p}, \{b_j - B_j s\}_{j=m+1}^{q}, \{f_j + F_j r - F_j s\}_{j=1}^{\tilde{Q}}, \{d_j + D_j\}_{j=1}^{M})}$$

lie to the right of $L_2$. In all of the subsequent applications, we shall simply take straight lines.

We introduce the chi-function (key function) for $z > 0$ (ultimately $\Re z > 0$) by

$$\mathrm{X}(z) = \frac{1}{2\pi i} \int_{L_1} \Delta(s)\chi(s) z^{-s} ds.$$

**Lemma 1 ([17, Theorem 1]).** *For* $\Re z > 0$ *we have a form of the modular relation equivalent to* (3.1). *If we can take* $\gamma_1 > \sigma_\phi$

$$\mathrm{X}(z) = \sum_{k=1}^{\infty} \alpha_k H_{p,q}^{m,n}\left( z\lambda_k \,\middle|\, \begin{matrix} \{(1-a_j, A_j)\}_{j=1}^{n}, \{(a_j, A_j)\}_{j=n+1}^{p} \\ \{(b_j, B_j)\}_{j=1}^{m}, \{(1-b_j, B_j)\}_{j=m+1}^{q} \end{matrix} \right),$$

*and if we can take* $\gamma_2 < r - \sigma_\psi$,

$$\mathrm{X}(z) = z^{-r} \sum_{k=1}^{\infty} \beta_k H^{n+\tilde{N}, m+P}_{q+P+\tilde{Q},\, p+M+\tilde{N}}$$

$$\times \left( \frac{\mu_k}{z} \middle| \begin{matrix} \{(1-b_j - B_j r, B_j)\}_{j=1}^m, \{(1-c_j - C_j r, C_j)\}_{j=1}^P, \\ \{(a_j - A_j r, A_j)\}_{j=1}^n, \{(e_j, E_j)\}_{j=1}^{\tilde{N}}, \end{matrix} \right.$$

$$\left. \begin{matrix} \{(b_j - B_j r, B_j)\}_{j=m+1}^q, \{(f_j, F_j)\}_{j=1}^{\tilde{Q}} \\ \{(1-a_j - A_j r, A_j)\}_{j=n+1}^p, \{(1-d_j - D_j r, D_j)\}_{j=1}^M \end{matrix} \right)$$

$$+ \sum_{k=1}^{L} \mathrm{Res}(\Delta(s)\chi(s) z^{-s}, s = s_k).$$

We also use the Meijer $G$-function which is defined as the special case of Fox $H$-function obtained by setting equal to 1 all coefficients denoted by capital letters in the definiton of the $H$-function:

$$G^{m,n}_{p,q}\left( z \middle| \begin{matrix} \{1-a_j\}_{j=1}^n, \{a_j\}_{j=n+1}^p \\ \{b_j\}_{j=1}^m, \{1-b_j\}_{j=m+1}^q \end{matrix} \right) = H^{m,n}_{p,q}\left( z \middle| \begin{matrix} \{(1-a_j, 1)\}_{j=1}^n, \{(a_j, 1)\}_{j=n+1}^p \\ \{(b_j, 1)\}_{j=1}^m, \{(1-b_j, 1)\}_{j=m+1}^q \end{matrix} \right).$$

*Proof of Theorem.* Since (1.10) is equivalent to ($\sigma < 0$)

$$\zeta(s,\alpha) = \frac{2\Gamma(1-s)}{(2\pi)^{1-s}} \left( \sin\frac{\pi}{2}s \sum_{n=1}^{\infty} \frac{\cos 2\pi\alpha n}{n^{1-s}} + \cos\frac{\pi}{2}s \sum_{n=1}^{\infty} \frac{\sin 2\pi\alpha n}{n^{1-s}} \right)$$

$$= \frac{2\pi}{(2\pi)^{1-s}\Gamma(s)\sin\pi s} \left( \sin\frac{\pi}{2}s \sum_{n=1}^{\infty} \frac{\cos 2\pi\alpha n}{n^{1-s}} + \cos\frac{\pi}{2}s \sum_{n=1}^{\infty} \frac{\sin 2\pi\alpha n}{n^{1-s}} \right),$$

we shall prove (1.10) in the form

$$\Gamma(s)\zeta(s,\alpha) = \frac{(2\pi)^s}{2} \left( \frac{1}{\cos\frac{\pi}{2}s} \sum_{n=1}^{\infty} \frac{\cos 2\pi\alpha n}{n^{1-s}} + \frac{1}{\sin\frac{\pi}{2}s} \sum_{n=1}^{\infty} \frac{\sin 2\pi\alpha n}{n^{1-s}} \right). \tag{3.2}$$

We use Lemma 1 with the data ($w$ being the integral variable)

$$\varphi(w) = \psi(w) = \pi^{-\frac{w}{2}}\zeta(w) = \sum_{k=1}^{\infty} \frac{1}{(\sqrt{\pi}k)^w},$$

so that $\sigma_\varphi = \sigma_\psi = 1$, $r = 1$, $\alpha_k = \beta_k = 1$, $\lambda_k = \mu_k = \sqrt{\pi}k$,

$$G(w) = \frac{\Gamma\left(\frac{1}{2}(1-w)\right)}{\Gamma\left(\frac{1}{2}w\right)},$$

so that $\tilde{N} = 1$, $E_1 = \frac{1}{2}$, $e_1 = 0$, $P = 0$, $M = 1$, $D_1 = \frac{1}{2}$, $d_1 = 0$, $\tilde{Q} = 0$;

$$\Delta(w) = \Gamma(s-w)\Gamma(w)$$

with

$$m = 1, \quad B_1 = 1, \quad b_1 = 0, \quad q = 1,$$
$$n = 1, \quad A_1 = 1, \quad a_1 = s, \quad p = 1.$$

Then for $\sigma > 1$, Lemma 1 gives

$$\begin{aligned} X(z) &= \sum_{k=1}^{\infty} \alpha_k H_{1,1}^{1,1}\left(z\lambda_k \,\middle|\, \begin{matrix} (1-s, 1) \\ (0, 1) \end{matrix}\right) \\ &= \sum_{k=1}^{\infty} G_{1,1}^{1,1}\left(\sqrt{\pi} zk \,\middle|\, \begin{matrix} 1-s \\ 0 \end{matrix}\right) \\ &= \sum_{k=1}^{\infty} \Gamma(s) \frac{1}{(\sqrt{\pi} zk + 1)^s} \end{aligned}$$

in view of the formula

$$G_{1,1}^{1,1}\left(z \,\middle|\, \begin{matrix} 1-a \\ b \end{matrix}\right) = \Gamma(a+b) \frac{z^b}{(z+1)^{a+b}}.$$

Hence

$$X(z) = \frac{\Gamma(s)}{(\sqrt{\pi} z)^s} \sum_{k=1}^{\infty} \frac{1}{\left(k + \frac{1}{\sqrt{\pi} z}\right)^s} = \alpha^s \Gamma(s)(\zeta(s, \alpha) - \alpha^{-s}), \tag{3.3}$$

where we put

$$\alpha = \frac{1}{\sqrt{\pi} z}. \tag{3.4}$$

On the other hand, for $-1 < \sigma < 0$, Lemma 1 gives

$$\begin{aligned} X(z) &= z^{-1} \sum_{k=1}^{\infty} \beta_k H_{1,3}^{2,1}\left(\frac{\mu_k}{z} \,\middle|\, \begin{matrix} (0, 1) \\ (s-1, 1), \left(0, \frac{1}{2}\right), \left(\frac{1}{2}, \frac{1}{2}\right) \end{matrix}\right) + P(z) \\ &= z^{-1} \sum_{k=1}^{\infty} f\left(\frac{\mu_k}{z}\right) + P(z), \end{aligned} \tag{3.5}$$

where

$$\begin{aligned} f(Y) &= \frac{1}{2\pi i} \int_{(\gamma_2)} \frac{\Gamma(s-1+w)\Gamma\left(\frac{1}{2}w\right)\Gamma(1-w)}{\Gamma\left(\frac{1-w}{2}\right)} Y^{-w} dw \\ &= \frac{1}{2\pi i} \int_{(c)} \frac{\Gamma(s-w)\Gamma(w)\Gamma\left(\frac{1-w}{2}\right)}{\Gamma\left(\frac{w}{2}\right)} Y^{w-1} dw, \end{aligned} \tag{3.6}$$

$c = 1 - \gamma_2$ and $1 < c\,(< 2 + \sigma)$ and

$$P(z) = \sum_{w=0,1} \mathrm{Res}(\Gamma(s-w)\Gamma(w)\pi^{-\frac{w}{2}}\zeta(w)z^{-w}) = -\frac{1}{2}\Gamma(s) + \Gamma(s-1)x. \quad (3.7)$$

We transform $f(Y)$ as follows. Since

$$\frac{\Gamma(\frac{1-w}{2})}{\Gamma(\frac{w}{2})} = \Gamma\left(\frac{1-w}{2}\right)\Gamma\left(1-\frac{w}{2}\right)\frac{\sin\frac{\pi w}{2}}{\pi}$$

$$= \Gamma(1-w)2^{w}\pi^{-\frac{1}{2}}\sin\frac{\pi w}{2},$$

it follows that

$$f(Y) = \frac{1}{2i\sqrt{\pi Y}}\frac{1}{2\pi i}\int_{(c)} \Gamma(s-w)\Gamma(w)\Gamma(1-w)(e^{\frac{\pi i}{2}w} - e^{-\frac{\pi i}{2}w})(2Y)^{w}dw$$

$$= \frac{1}{2i\sqrt{\pi Y}}\left\{G_{1,2}^{2,1}\left(2iY\,\middle|\,\begin{matrix}1\\ s,\,1\end{matrix}\right) - G_{1,2}^{2,1}\left(-2iY\,\middle|\,\begin{matrix}1\\ s,\,1\end{matrix}\right)\right\}.$$

Now we apply

$$G_{1,2}^{2,1}\left(z\,\middle|\,\begin{matrix}1\\ s,\,1\end{matrix}\right) = \Gamma(s)z^{s/2}e^{z/2}W_{-s/2,s/2-1/2}(z) \quad \text{cf. [9, p. 216 (8)]}$$

$$= \Gamma(s)z^{s}\left\{\Gamma(1-s)\Phi(s;s;z) + \frac{\Gamma(s-1)}{\Gamma(s)}z^{1-s}\Phi(1;2-s;z)\right\}$$

cf. [9, p. 264 (2) and p. 257 (7)]

$$= \Gamma(s)\Gamma(1-s)z^{s}e^{z} + \Gamma(s-1)z\,\Phi(1;2-s;z),$$

where $\Phi(a;c;z)$ is the confluent hypergeometric function ${}_1F_1(a;c;z)$. It is easy to see that

$$z\,\Phi(1;2-s;z) = (1-s)\{\Phi(1;1-s;z) - 1\},$$

whence we have

$$f(Y) = \frac{1}{2i\sqrt{\pi Y}}\frac{\pi}{\sin\pi s}((2iY)^{s}e^{2iY} - (-2iY)^{s}e^{-2iY})$$

$$- \frac{1}{2i\sqrt{\pi Y}}\Gamma(s)(\Phi(1;1-z;2iY) - \Phi(1;1-s;-2iY)).$$

The first term is

$$\frac{1}{2i\sqrt{\pi Y}}\frac{\pi}{\sin\pi s}(2Y)^{s}\left(e^{(2Y+\frac{\pi}{2}s)i} - e^{-(2Y+\frac{\pi}{2}s)i}\right)$$

$$= \frac{2\sqrt{\pi}}{\sin\pi s}(2Y)^{s-1}\sin\left(2Y+\frac{\pi}{2}s\right)$$

$$= \sqrt{\pi}(2Y)^{s-1}\left(\frac{\sin 2Y}{\sin\frac{\pi s}{2}} + \frac{\cos 2Y}{\cos\frac{\pi s}{2}}\right).$$

Hence

$$f(Y) = \sqrt{\pi}(2Y)^{s-1}\left(\frac{\sin 2Y}{\sin\frac{\pi}{2}s} + \frac{\cos 2Y}{\cos\frac{\pi}{2}s}\right)$$

$$- \frac{1}{2i\sqrt{\pi}Y}\Gamma(s)(\Phi(1; 1-s; 2iY) - \Phi(1; 1-s; -2iY)). \tag{3.8}$$

Substituting (3.7) and (3.8) into (3.5), we obtain

$$X(z) = z^{-1}\sum_{k=1}^{\infty}\sqrt{\pi}\left(\frac{2\pi k}{\sqrt{\pi}z}\right)^{s-1}\left(\frac{\sin\frac{2\pi k}{\sqrt{\pi}z}}{\sin\frac{\pi}{2}s} + \frac{\cos\frac{2\pi k}{\sqrt{\pi}z}}{\cos\frac{\pi}{2}s}\right)$$

$$- \frac{\Gamma(s)}{2\pi i}\sum_{k=1}^{\infty}\frac{1}{k}\left(\Phi\left(1; 1-s; \frac{2\pi ik}{\sqrt{\pi}z}\right) - \Phi\left(1; 1-s; -\frac{2\pi ik}{\sqrt{\pi}z}\right)\right)$$

$$- \frac{1}{2}\Gamma(s) + \Gamma(s-1)\alpha. \tag{3.9}$$

Comparing (3.3) and (3.9), we obtain for $\sigma < 0$

$$\alpha^s\Gamma(s)\zeta(s,\alpha) = \frac{1}{2}(2\pi\alpha)^s\sum_{k=1}^{\infty}\frac{1}{k^{1-s}}\left(\frac{\sin 2\pi\alpha k}{\sin\frac{\pi}{2}s} + \frac{\cos 2\pi\alpha k}{\cos\frac{\pi}{2}s}\right)$$

$$- \frac{1}{2\pi i}\Gamma(s)\sum_{k=1}^{\infty}\frac{1}{k}g(2\pi\alpha k) + \frac{\Gamma(s)}{2} + \Gamma(s-1)\alpha, \tag{3.10}$$

where

$$g(z) = \Phi(1; 1-s; iz) - \Phi(1; 1-s; -iz).$$

It remains to prove that

$$-\frac{1}{2\pi i}\sum_{k=1}^{\infty}\frac{1}{k}g(2\pi\alpha k) + \frac{1}{2} + \frac{\alpha}{s-1} = 0. \tag{3.11}$$

For this we transform $g(z)$ by the formula

$$\Phi(a; c; z) = \frac{2^{1-c}e^{\frac{z}{2}}\Gamma(c)}{\Gamma(a)\Gamma(c-a)}\int_{-1}^{1}e^{\frac{zt}{2}}(1-t)^{c-a-1}(1+t)^{a-1}dt$$

($\operatorname{Re} c > \operatorname{Re} a > 1$), with $a = 1$ or $c = 1 - s$:

$$\Phi(1; 1-s; iz) = \frac{2^s\Gamma(1-s)}{\Gamma(-s)}e^{\frac{iz}{2}}\int_{-1}^{1}e^{\frac{izt}{2}}(1-t)^{-s-1}dt. \tag{3.12}$$

By this expression we have

$$g(z) = -s2^s\int_{-1}^{1}(e^{\frac{iz}{2}(1+t)} - e^{-\frac{iz}{2}(1+t)})(1-t)^{-s-1}dt$$

$$= -s2^{s+1}i\int_{-1}^{1}\sin\left(\frac{z}{2}(1+t)\right)(1-t)^{-s-1}dt.$$

Hence the first term in (3.11) is

$$\frac{2^s}{\pi}s\sum_{k=1}^{\infty}\frac{1}{k}\int_{-1}^{1}\sin(\pi\alpha k(1+t))(1-t)^{-s-1}dt$$

$$=\frac{2^s}{\pi}s\int_{-1}^{1}\sum_{k=1}^{\infty}\frac{\sin(\pi\alpha k(1+t))}{k}(1-t)^{-s-1}dt.$$

Now we recall the special case of (2.16) mentioned at the end of section 1.

$$\sum_{k=1}^{\infty}\frac{\sin kx}{k}=\frac{1}{2}(\pi-x)\quad \text{for } 0<x<2\pi. \tag{3.13}$$

(Note that the parameter $\alpha$ is implicitly assumed that $0<\alpha\leq 1$.) Substituting (3.13), we see that the integral is

$$\frac{\pi}{2}\int_{-1}^{1}(1-\alpha(1+t))(1-t)^{-s-1}dt=-\frac{\pi}{2}\left(\frac{1-2\alpha}{s}2^{-s}-\frac{\alpha}{1-s}2^{1-s}\right),$$

in view of $\sigma<0$. Hence the first term becomes

$$-\frac{s}{2}\left(\frac{1-2\alpha}{s}-\frac{2\alpha}{1-s}\right)=-\frac{1}{2}\left(1-2\alpha-\frac{2\alpha s}{1-s}\right),$$

which cancels the second and third terms, and (3.11) follows. This completes the proof of Theorem 3. □

## Acknowledgements

The main body of this paper was written while the third and fourth authors were staying at National University of Singapore. They would like to thank Professor H. H. Chan for his hospitality.

## References

[1] T. M. Apostol, Generalized Dedekind sums and transformatlon formulae of certain Lambert series, *Duke Math. J.* **17** (1950) 147–157

[2] T. M. Apostol, A short proof of Sho Iseki's functional equation, *Proc. Amer. Math. Soc.* **15** (1964) 618–622

[3] A. Basu, Alternate proof of Sho Iseki's functional euqation, *Ramanujan J.* **3** (1999) 297–301

[4] B. C. Berndt, Generalized Dedekind eta-functions and generalized Dedekind sums, *Trans. Amer. Math. Soc.* **178** (1973) 495–508

[5] B. C. Berndt, Modular transformations and generalizations of several formulae of Ramanujan, *Rocky Mount. J. Math.* **7** (1977) 147–189

[6] R. Campbell, Les intégrales Eulériennes et leurs applications, Etude approfondie de la fonction gamma, Dunod, Paris, 1966

[7] C. Deninger, On the analogue of the formula of Chowla and Selberg for real quadratic fields, *J. Reine Angew. Math.* **351** (1984) 172–191

[8] J. Elstrodt, A very simple proof of the eta transformatlon formula, to appear

[9] A. Erdélyi, et al. (Bateman Manuscript Project), Higher transcendental functions I, Mcgraw-Hill, 1953

[10] L. J. Goldstein and P. de la Torre, On the transformation formula of $\log \eta(\tau)$, *Duke Math. J.* **41** (1974) 291–297

[11] E. Grosswald, Comments on some formulae of Ramanujan, *Acta Arith.* **21** (1972) 25–34

[12] S. Iseki, The transformation formula for the Dedekind modular function and related functional equations, *Duke Math. J.* **24** (1957) 653–662

[13] S. Iseki, A proof of a functional euqation related to the theory of partitions, *Proc. Amer. Math. Soc.* **12** (1961) 502–505

[14] S. Kanemitsu, The use of the cotangent function in number theory, Proc. SASTRA Conf., to appear

[15] S. Kanemitsu, M. Katsurada and M. Yoshimoto, On the Hurwitz-Lerch zeta-function, *Aeq. Math.* **59** (2000) 1–19

[16] S. Kanemitsu, Y. Tanigawa, H. Tsukada and M. Yoshimoto, Some aspects of the modular relation, Number Theory: Tradition and Modernization (ed. W. Zhang and Y. Tanigawa), DEVM 15, Springer, (2006) 103–118

[17] S. Kanemitsu, Y. Tanigawa and H. Tsukada, Some number theoretic applications of a general modular relation, *Int. J. Number Theory* **2** (2006) 599–615

[18] S. Kanemitsu, Y. Tanigawa and H. Tsukada, Contributions to the theory of zeta-functions – the modular relation supremacy, in preparation

[19] N. S. Koshlyakov, Investigation of some questions of the analytic theory of a rational and quadratic field I (Russian), *Izv. Akad. Nauk SSSR Ser. Mat.* **18** (1954) 113–144

[20] A. Laurinčikas and R. Garunkštis, The Lerch zeta-function, Kluwer, Dordrecht etc. 2002

[21] H.-L. Li, M. Hashimoto and S. Kanemitsu, Structural elucidation of Eisenstein's formula, to appear

[22] M. Mikolás, Über gewisse Lambertsche Reihen I: Verallgemeinerung der Modulfunktion $\eta(\tau)$ und ihrer Dedekindschen Transformationsformel, *Math. Z.* **68** (1957) 100–110

[23] J. Milnor, On polylogarithms, Hurwitz zeta functions, and the Kubert identities, *Enseign. Math. (2)* **29** (1983) 281–322

[24] H. Rademacher, Zur Theorie der Modulfunktionen, *J. Reine Angew. Math.* **167** (1931) 312–336 = Collected Papers of Hans Rademacher, Vol. 1, MIT Press, Cambridge etc. 1974, 652–676

[25] R. Sczech, Ein einfacher Beweis der Transformationsformel für $\log \eta(z)$, *Math. Ann.* **237** (1978) 161–166

[26] E. C. Titchmarsh, The theory of the Riemann zeta-function, Clarendon. Press, Oxford 1951; revised by D. R. Heath-Brown, 1986

[27] K. Wang, Exponential sums of Lerch's zeta functions, *Proc. Amer. Math. Soc.* **95** (1985) 11–15

[28] X.-Y. Wang, S. Kanemitsu and Y. Tanigawa, An application of the Euler digamma function, preprint.

[29] A. Weil, Elliptic functions according to Eisenstein and Kronecker, Springer, Berlin-Heidelberg-New York, 1976

Proc. Int. Conf. – Number Theory and Discrete Geometry No. 6, 2008, pp. 35–41.

# Some Unfinished Tasks

R. P. Bambah

Department of Mathematics, Panjab University, Chandigarh 160 014, India
e-mail:bambahrp@rediffmail.com

Here I shall describe some of the problems in which I have been interested and which have defied solution despite attempts by several mathematicians.

## 1. Sums of two squares

Let $s_1, s_2, \ldots$ be an increasing sequence of positive integers of the type $a^2+b^2$, where $a, b$ are integers.

Chowla and I used a straight forward construction to prove that

$$s_{n+1} - s_n = O(s_n)^{1/4}. \tag{1}$$

More precisely if $c > 2\sqrt{2}$, there exist integers $a, b$ such that

$$x < a^2 + b^2 < x + cx^{1/4}, \tag{2}$$

for all $x$ large enough.

We conjectured that the exponent $1/4$ can be replaced by any $\epsilon > 0$. However, in spite of the efforts of many mathematicians, including L. J. Mordell, J. E. Littlewood and P. Erdős, no improvement in (1) has been obtained so far. Slight improvements in (2) and generalizations have been made by Majumdar (1950), S. Uchiyama (1965), P. H. Diananda (1966), and L. J. Mordell (1969). Hooley proved a related result.

## 2. Ramanujan's function $\tau(n)$

S. Ramanujan defined the function $\tau(n)$ by

$$\sum_{1}^{\infty} \tau(n)x^n = x\left(\prod_{r=1}^{\infty}(1-x^r)\right)^{24}.$$

He proved, among other things,

$$\tau(n) \equiv n\sigma(n) \pmod 5$$

and

$$\tau(n) \equiv \sigma_{11}(n) \pmod{691},$$

where

$$\sigma_k(n) = \sum_{d|n} d^k = \text{sum of the } k^{\text{th}} \text{ powers of positive divisors of n.}$$

Various mathematicians, H. Gupta, K. G. Ramanathan, D. B. Lahiri, G. L. Watson, R. J. Wilton and others proved further results of this type. So did the author and S. Chowla for powers of 2, 3, 5, 7, 23, 691 and later for others.

D. H. Lehmer asked if $\tau(n) = 0$ for any $n$. He proved that the smallest n with $\tau(n) = 0$ must be a prime. From results of this type he deduced that $\tau(p) \neq 0$ if $p < 3,316,799$. This result can be improved by using later results. However, the question whether $\tau(n) = 0$ for any n is still open.

On the other hand, S. Chowla (improving on earlier results) has proved that if $a, b, c, d, e, f$ are non-negative integers then

$$\tau(n) \equiv 0 \ (\text{mod} 2^a 3^b 5^c 7^d 23^e 691^f)$$

for almost all $n$.

## 3. Lattice coverings

Let $\mathbb{R}^n$ be the $n$-dimensional Euclidean space. Let $K \subset \mathbb{R}^n$. Let $\Lambda$ be a lattice of determinant $d(\Lambda)$. If the sets $K + A$, $A \in \Lambda$ (where $K + A = \{x + A : x \in K\}$ is the translate of $K$ through $A$) cover $\mathbb{R}^n$ i.e. if $\mathbb{R}^n \subset \bigcup_{A\in\Lambda} K + A$ then $(K, \Lambda)$ is called a lattice covering by $K$. For simplicity, suppose that $K$ has volume $V(K)$. Then $\theta(K, \Lambda) = V(K)/d(\Lambda)$ is the density of the covering $(K, \Lambda)$ and $\theta(K) = \inf\{\theta(K, \Lambda) : (K, \Lambda) \text{ lattice covering}\}$ is called the density of the thinnest lattice covering by $K$.

## 4. General coverings

If $K \subset \mathbb{R}^n$ and $\sum \subset \mathbb{R}^n$ are any sets, we say that $(K, \sum)$ is a covering by $K$ if $\mathbb{R}^n \subset \bigcup_{A\in\sum}(K + A)$. We can define $\theta(K, \sum)$, the density of this covering as follows:

Let $\mathcal{B} : |x_i| \leq l, 1 \leq i \leq n$. Let $N(\mathcal{B})$ = the number of $K + A$ such that $K + A \cap \mathcal{B} \neq \emptyset$, $A \in \sum$, $\theta(K, \sum, \mathcal{B}) = N(\mathcal{B})V(K)/V(\mathcal{B})$, then

$$\theta(K, \sum) = \liminf_{l\to\infty} \theta(K, \sum, \mathcal{B})$$

is called the covering density of the covering $(K, \sum)$.

$\theta^*(K) = \inf\{\theta(K, \sum) : (K, \sum) \text{ covering}\}$ is called the density of the thinnest covering of $K$.

Clearly

$$\theta(K) \geq \theta^*(K) \geq 1.$$

If $K$ is an $n$-dimensional sphere then we write $\theta_n$ for $\theta(K)$ and $\theta_n^*$ for $\theta^*(K)$.

I was introduced to these by H. Davenport in 1950.

The first explicit results in literature appear to be due to Kershner (1939):

$$\theta_2 = \theta_2^* = \frac{2\pi}{3\sqrt{3}}.$$

This result was extended in early 1950's by L. Fejes Tóth and by R. P. Bambah and C. A. Rogers to prove:

If $K$ is a symmetrical convex domain in $\mathbb{R}^2$ of area $a(K)$, then

$$\theta(K) = \theta^*(K) = \frac{a(K)}{h_s(K)},$$

where $h_s(K)$ is the area of the largest symmetrical hexagon (polygon with atmost 6 sides) contained in $K$ and $\theta(K) = \theta^*(K) \geq 2\pi/(3\sqrt{3})$. Other proofs and extensions were given by R. P. Bambah and C. A. Rogers as also by R. P. Bambah, C. A. Rogers and H. Zassenhaus and R. P. Bambah and A. C. Woods, among others.

It is an open question whether $\theta(K) = \theta^*(K)$ for convex domains without a centre.

That the result does not hold for the larger class of symmetrical star domains was proved by R. P. Bambah, V. C. Dumir and R. J. Hans-Gill (1977). (Earlier result for higher dimensions is due to S. K. Stein (1972)).

## 5. Bounds for $\theta(K), \theta^*(K), \theta_n, \theta_n^*$

If $K$ is a convex body with a centre in $\mathbb{R}^n$, Hlawka (1949) showed $\theta(K) \leq n^n$. This was improved by C. A. Rogers in 1950 to $\theta(K) \leq 3^{n-1}$. R. P. Bambah and K. F. Roth (1951) showed that if $K$ has enough symmetries, $\theta(K) \leq C_n \sim \sqrt{\frac{\pi}{54n}}e^n$ (D. B. Sawyer showed without any symmetry restriction on $K$ that $\theta(K) \leq \frac{n^n}{n!} \sim \frac{e^n}{\sqrt{2\pi n}}$ for all convex $K$, best possible if $\Lambda$'s are restricted to a special type). However earlier, after the work of W. M. Schmidt and C. A. Rogers one has

$$\theta(K) \leq n^{\log_2 \log n + \epsilon}. \tag{3}$$

For $\theta^*(K)$, C. A. Rogers showed

$$\theta^*(K) \leq n \log n + o(n \log n). \tag{4}$$

(Notice the gap between the bounds on $\theta(K)$ and $\theta^*(K)$).

For $\theta_n, \theta_n^*$, C. A. Rogers showed that

$$\theta_n \leq cn(\log n)^{\frac{1}{2}\log_2 2\pi e} \tag{5}$$

which is an improvement on earlier results of H. Davenport etc. of the type $\theta_n \leq c^n$.

On the other hand H. Davenport and R. P. Bambah (1952) showed

$$\theta_n \geq \frac{4}{3} - \epsilon_n, \text{ where } \epsilon_n \to 0 \text{ as } n \to \infty$$

and P. Erdős and C. A. Rogers (1953) showed

$$\theta_n^* \geq \frac{16}{15} - \epsilon_n, \quad \text{where } \epsilon_n \to 0 \text{ as } n \to \infty.$$

However, soon afterwards Coxeter, Few and Rogers (1959) showed

$$\theta_n \geq \theta_n^* \geq \tau_n \sim \frac{n}{e\sqrt{e}}. \tag{6}$$

(Notice that (4), (5) and (6) are comparable.)

## 6. Evaluation of $\theta_n, \theta_n^*$

As talked earlier, Kershner (1939) showed

$$\theta_2 = \theta_2^* = \frac{2\pi}{3\sqrt{3}}.$$

(Perhaps $\theta_2 = \frac{2\pi}{3\sqrt{3}}$ was known to Dirichlet.)

In 1954, the author proved

$$\theta_3 = \frac{5\sqrt{5}\pi}{24}.$$

Other proofs were given by E. S. Barnes (1956) and L. Few (1956). $\theta_4 = 2\pi^2/(5\sqrt{5})$ was proved by Delone and Ryskov and by T. J. Dickson in 1960's.

$\theta_5$ was determined by Ryskov and Baranovsky in 1970's. For $n \geq 6$, the problem of determination of $\theta_n$ is still open.

Regarding $\theta_n^*$, as we have already stated $\theta_2^*$ is due to Kershner. It is not known whether $\theta_3^* = \theta_3 = 5\sqrt{5}\pi/24 = 1.464\ldots$, as it happens in the case of packings of spheres, Kepler's conjecture having been proved by T. C. Hales a few years earlier. Coxeter, Few and Rogers (1959) showed that $\theta_3^* \geq 1.431$, which has been perhaps improved a little later. If $\Sigma$ is union of a lattice $\Lambda$ and a translate of $\Lambda$ and $\Sigma$ is covering set for the three-dimensional sphere K, then

$$\theta(K, \Sigma) \geq \theta_3,$$

was proved by R. P. Bambah and A. C. Woods (1971). But so far, such a result has not been proved for more than two translates of a lattice. Proving it for arbitrarily finite number of translates would lead to $\theta_3 = \theta_3^*$.

No relation between $\theta_n, \theta_{n+1}$ is known so far. Method used by me to get lower bound for $\theta_4$ from the knowledge of $\theta_3$ would perhaps give some weak results.

## 7. Finite covering

Improving earlier results of Bambah & Rogers, and Bambah, Rogers & Zassenhaus, the following result was proved by A. C. Woods & author:

Let $K$ be a closed convex domain with centre. Let

$$A_0, A_1, \dots, A_n = A_0, A_{n+1}, \dots, A_{n+m}, \; m \geq 0$$

be points such that $A_0, \dots, A_n$ is the boundary of a closed convex domain $\pi$ of area $a(\pi)$, which is covered by the union sets $K + A_i$, $1 \leq i \leq n + m$. Then

$$a(\pi) \leq (2m + n - 2)t(K),$$

where $t(K)$ is the area of the largest triangle contained in $K$. (Prof. Wills will perhaps talks about related results.)

## 8. Minkowski's Conjecture and related results

A conjecture attributed to Minkowski states:

Let $L_i = \sum_{j=1}^{n} a_{ij}x_j$, $a_{ij} \in \mathbb{R}$ be $n$ real linear forms with $\Delta = \det(a_{ij}) \neq 0$. Then given real $c_1, \dots c_n$, there exist integers $x_1, x_2, \dots, x_n$ satisfying

$$|\Pi(L_i + c_i)| \leq |\Delta|/2^n.$$

(The result, if true, is best possible). Minkowski (1899) proved the case $n = 2$. Several other proofs have since been given.

For $n = 3$, Remak (1923) gave a proof in two papers covering 62 pages.

Davenport (1939) gave a four page proof. Other proofs were given by Birch and Swinnerton-Dyer (1956) and Narzulliev (1968). For $n = 4$, Dyson gave a proof in 1948. Like some earlier proof for $n = 2, 3$, Dyson's proof consisted of two parts:

**Part I:** Every lattice $\Lambda$ can be brought to a "well-rounded" form by suitable automorphism of the region $K : |\Pi x_i| \leq 1$,

**Part II:** Every "well-rounded" lattice of det 1 is a covering lattice for the sphere $|x| \leq \frac{\sqrt{n}}{4}$ and hence region $|\Pi x_i| \leq \frac{1}{2^n}$.

Dyson's proof of Part II is elementary though complicated. Simpler proofs were given later by Cleaver (1965) and A. C. Woods (1965).

Dyson's proof of Part I used strong tools from Algebraic Topology. Using a result of Birch and Swinnerton-Dyer, for the purpose of proving Minkowski's conjecture, one can disregard lattices which have points on or near the coordinate planes. For the others, Bambah and Woods (1974) gave a proof, which is elementary. Skubenko (1973) announced another proof which was published later.

For Part II, A. C. Woods gave proofs for $n = 5$, 6 in 1965 and 1972 respectively.

For Part I, Skubenko announced a proof for $n = 5$ which he published later. However, the editor of the journal remarked that the proof was somewhat incomplete and the paper may be regarded as an initial publication of this notable result. Bambah and Woods, following Skubenko's plan, gave in 1980 a complete proof, which differs in some details from Skubenko's proof. This completed the proof of Minkowski's conjecture for $n = 5$.

I understand Curtis McMullen has proved Part I stated above for all $n$. (Jl. American Math. Soc. 18 (2005) 711–734). The problem is open for $n \geq 7$.*

*R.J. Hans-Gill, Madhu Raka and Ranjeet Sehmi have proved Part II for $n = 7$ in 2007.

Chebotoroff (1934) proved the weaker result that

$$|\Pi(L_i + c_i)| \leq \frac{|\Delta|}{(\sqrt{2})^n}$$

has a solution. Later, Davenport, Mordell, Woods, Bombieri, Gruber gave improvements replacing $(\sqrt{2})^n$ by larger numbers $(\sqrt{2})^n \alpha_n$, where $\alpha_n = O(1)$. A qualitative improvement came from Skubenko's school (Skubenko, Narzullaev, Mukhsinov, etc.). Best result, as far as I know, is of the type

$$\alpha_n \cong O\left(\sqrt{\frac{n}{\log n}}\right).$$

Chalk (1947) showed the solvability of the inequalities

$$L_1 + c_1 > 0, \ldots, L_n + c_n > 0, \ \Pi(L_i + c_i) \leq |\Delta|.$$

This result is clearly best possible.

Cole(1952) showed that

$$L_1 + c_1 > 0, \ldots, L_{n-1} + c_{n-1} > 0, \ |\Pi(L_i + c_i)| \leq |\Delta|/2$$

are solvable.

This suggests the conjecture that for $0 \leq r \leq n$, we have integral solutions for

$$L_1 + c_1 > 0, \ldots, L_r + c_r > 0, \ |\Pi(L_i + c_i)| \leq \frac{|\Delta|}{2^{n-r}}.$$

For $n = 2$, the conjecture follows from the results of Minkowski ($r = 0$), Cole ($r = 1$) and Chalk ($r = 2$). For $n = 3$, the result follows for $r = 0$ (Davenport-Remak), $r = 3$ (Chalk), $r = 2$ (Cole). The missing case $r = 1$ was settled by Bambah and Woods (1977). For $n \geq 4$, the conjecture is open except for the cases $r = n, n - 1$ for all $n$ and the case $r = 0, \ 4 \leq n \leq 6$, as mentioned earlier.

## 9. Non-homogeneous indefinite quadratic forms

Minkowski's result for $n = 2$ above can be easily interpreted as one on non-homogeneous binary quadratic forms. This can be generalized to a problem of quadratic form in $n$ variables. As a result of the work of Davenport, Heilbronn, Blaney, C. A. Rogers, D. M. E. Foster, B. J. Birch, G. L. Watson, Barnes, Dumir, Hans-Gill, Madhu Raka, Flahive, Jackson, Sehmi, Bambah, Aggarwal, Gupta and eventually the joint work of Dumir, Hans-Gill and Woods, the problem has been completely solved for the so called symmetric case. For the positive case, out of the infinitely many pairs $(r, s)$, $r$, $s$, positive integers, only one case $(r, s) = (1, 4)$ remains unsolved. I expect Madhu Raka and Hans-Gill to talk about this in detail.

Instead of listing all the references in Sections 3–9, we list only Bambah, Dumir and Hans-Gill [14] and Gruber and Lekkerkerker [15], where most references can be found.

## References for Sections 1 and 2


[1] R. P. Bambah, S. Chowla, On numbers which can be expressed as a sum of two squares. *Proc. Nat. Inst. Sci. India* **13** (1947) 101–103

[2] R. P. Bambah, S. Chowla, The residue of Ramanujan's function $\tau(n)$ to the modulus $2^8$. *J. London Math. Soc.* **22** (1947) 140–147

[3] S. Chowla, On a theorem of Walfisz. *J. London. Math. Soc.* **22** (1947) 136–140

[4] Diananda, Palahenedi Hewage, On numbers expressible as a weighted sum of powers. *Proc. Japan Acad.* **44** (1968) 890–894

[5] G. H. Hardy, Ramanujan Cambridge (1940) 165–167

[6] Hooley, Christopher, On the intervals between numbers that are sums of two squares. *Acta Math.* **127** (1971) 279–297

[7] D. H. Lehmer, The vanishing of Ramanujan's function $\tau(n)$. *Duke Math. J.* **14** (1947) 429–433

[8] Majumdar, N. Kulendra, On numbers which can be expressed by a given form. *Proc. Nat. Inst. Sci. India* **16** (1950) 99–100

[9] L. J. Mordell, On numbers which can be expressed as a sum of powers. 1969 Number Theory and Analysis (Papers in Honor of Edmund Landau) pp. 217–221 Plenum, New York

[10] S. Ramanujan, Collected Papers, Cambridge (1927) 142–146

[11] R. A. Rankin, Ramanujan's function $\tau(n)$, *Symposium on Theoretical Physics and Mathematics*, **10** (Inst. Math. Sci. Madras), (1970)

[12] Uchiyama, Saburo̧, On the distribution of integers representable as a sum of two $h$-th powers. *J. Fac. Sci. Hokkaido Univ. Ser. I* **18** (1964/1965) 124–127

[13] F. Van der Blij, The function $\tau(n)$ of S. Ramanujan (an expository lecture). *Math. Student* **18** (1950) 83–99


## References for Sections 3–9


[14] R. P. Bambah, V. C. Dumir and R. J. Hans-Gill, Non-homogeneous problems: conjectures of Minkowski and Watson, Number Theory 15–41, Trends in Mathematics, Birkhaüser Verlag, Basel (2000)

[15] P. Gruber, and C. G. Lekkerkerker, Geometry of Numbers, Second Edition, North Holland (1987)

Proc. Int. Conf. – Number Theory and Discrete Geometry No. 6, 2008, pp. 43–48.

# Incenter Iterations in the Plane and on the Sphere

A. Bezdek[1,2] and T. Bisztriczky[3]

[1]Department of Mathematics and Statistics, Auburn Univ., Auburn, AL
[2]Alfréd Rényi Institute of Mathematics, Budapest, Hungary
e-mail: bezdean@auburn.edu
[3] Department of Mathematics and Statistics, University of Calgary, Canada
e-mail: tbisztri@math.ucalgary.ca

**Abstract.** The central problem of this paper is the question of denseness of those planar point sets $\mathcal{P}$, that have the property that for every three noncollinear points in $\mathcal{P}$, a specific triangle center (incenter (IC), circumcenter ($CC$), orthocenter ($OC$) etc.) is also in the set $\mathcal{P}$. The Euclidean planar versions of the $IC$, $CC$ and $OC$ problems have been settled before. In this paper we give a second shorter proof for the $IC$ problem in the Euclidean plane with an approach that allows us to settle the hyperbolic and the spherical versions of the problem.

**Keywords.** Dense point set, incenter, iterated process.

**Subject Classification:** 52A35

## 1. Introduction

There are several results in the literature concerning iterative processes in the plane. A typical problem starts with the description of a geometric construction, which when applied to an initial point set generates larger point sets. The problem usually is to prove that repeated expansions lead to an everywhere dense point set.

For example, Ismailescu et al. [3] showed recently that starting with three noncollinear points, the repeated use of the construction "add the circumcenters (incenters resp.) of all nondegenerate triangles formed by the existing points" leads to dense point sets in the Euclidean plane (in the convex hull of the initial triangle resp.). Inspired by these results, G. Ambrus and A. Bezdek [1,2] solved the analogous questions concerning i) incenters in $\mathbb{E}^3$ and ii) orthocenters in $\mathbb{E}^2$.

In this paper, we give a second shorter proof for the incenter problem in the Euclidean plane and apply the method to the hyperbolic and spherical versions of the problem.

Let $\mathbb{E}^2$, $\mathcal{H}^2$ and $\mathcal{S}^2$ denote, respectively, the Euclidean plane, the hyperbolic plane and the sphere of radius one in $\mathbb{E}^3$. $\Delta(A, B, C)$ (or $\Delta$ in short) denotes the closed triangle in $\mathbb{E}^2$, $\mathcal{H}^2$ or $\mathcal{S}^2$ with vertices $A$, $B$ and $C$. We study the following iterative construction concerning incircles:

Let $S_0$ denote the set of vertices of a triangle $\Delta$. Let $i \geq 0$ and assume that $S_i$ has been defined. Then $S_{i+1}$ is the set containing $S_i$ and the set of incenters of all triangles with vertices in $S_i$. Finally let $S = \cup_{i=1}^{\infty} S_i$. We prove

The first author was supported by OTKA Grant T038397 and T043520, the second author was supported by a Natural Sciences and Engineering Research Council of Canada Discovery Grant.

**Theorem 1.1.** *$S$ is dense in $\Delta$; that is, the closure $\overline{S}$ of $S$ is $\Delta$.*

Theorem 1.1 answers affirmatively a question of D. Ismailescu concerning triangles in the Euclidean plane. In the proof Theorem 1.1, we assume that $\overline{S} \neq \Delta$ and seek a contradiction. Since $\overline{S} \neq \Delta$, there is a maximal closed circular disk $D$, whose center $d$ belongs to $\Delta$ and which is empty, meaning that $S \cap \text{int}D = \emptyset$.

If $d \in \text{int}\Delta$ then exactly two points, say, $x$ and $y$ of $\overline{S}$ belong to $\text{bd}(D)$. Indeed if there are at least three such points, then the incentre of the triangle spanned by them contradicts the emptiness of the circle $D$. On the other hand, if there is at most one such point, then an appropriate small enlargement of the circle $D$ (here small enlargement refers to moving the center by a small distance and then increasing the radius by a small number) contradicts the maximality of the circle $D$. By the same reasoning, we obtain that $x$ and $y$ are antipodal points of $D$.

In summary, it remains to seek a contradiction in the following two cases in each of $\mathbb{E}^2$, $\mathcal{H}^2$ and $\mathcal{S}^2$.

Case 1: $d \in \text{int}\Delta$, $\text{bd}D \cap \overline{S} = \{x, y\}$ and $x$ and $y$ are antipodal points of D.
Case 2: $d \in \text{bd}\Delta$ and $bdD \cap \text{int}\Delta \cap \overline{S} \neq \emptyset$.

## 2. Δ is a Euclidean triangle

First we show that Case 1 is not possible. We observe that Case 1 is depicted in Figure 1, which is oriented so that the diagonal $xy$ is vertical with vertex $x$ at the upper end. Let $U_{\text{left}}$ and $U_{\text{right}}$ denote the left and right half planes bounded by line $xy$.

For a small open $\epsilon$-neighbourhood of $x$, we say that a sequence of distinct points $\{p_n\}$ in this neighbourhood are converging to $x$ *horizontally from left (horizontally from right)* if all the points of this sequence belong to $U_{\text{left}}$ (all of them belong to $U_{\text{right}}$) and the angle sequence $\angle p_n xy$ has limit $\frac{\pi}{2}$. We define horizontal sequences at $y$ similarly. If for a sufficiently small $\epsilon$, the $\epsilon$-neighbourhoods of both $x$ and $y$ do not contain sequences in $S$ converging to $x$ or $y$ horizontally from left, then there is an appropriate small enlargement of circle $D$ which contradicts its maximality. The same may be said about the right direction. Thus in view of the obvious symmetries, we may assume that there are sequences $\{x_n\}$ and $\{y_n\}$ in $S$ converging horizontally

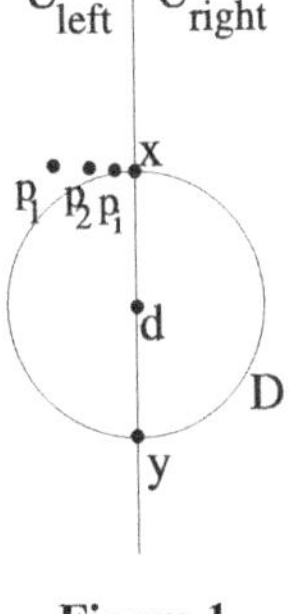


**Figure 1.**

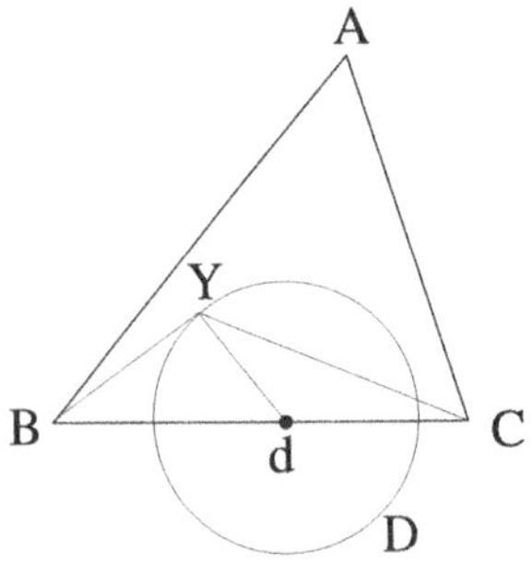


**Figure 2.**

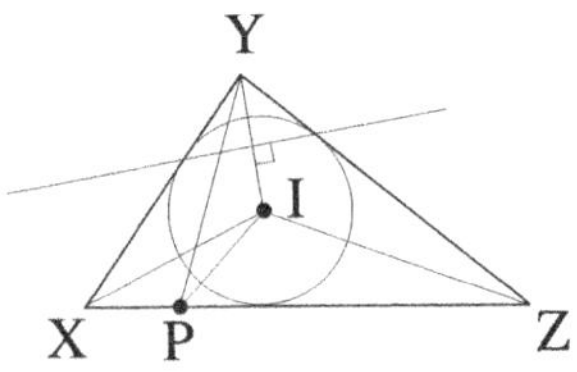


**Figure 3.**

from left to $x$ and to $y$, respectively. First we consider the triangles $\Delta(x, y, y_n)$. It is clear that for a $y_n$ sufficiently close to $y$, the angle $\angle xyy_n$ is close to $\frac{\pi}{2}$. Thus the incenter of $\Delta(x, y, y_n)$ is in int$D$ for $y_n$ sufficiently close to $y$. Then fixing such a $y_n$, a similar argument yields that for a $y_m$ sufficiently close to $y$, the incenter of $\Delta(x, y_m, y_n)$ is also in int$D$. We now fix also such a $y_m$. Finally, the preceding yield that the incenter of $\Delta(x_k, y_m, y_n)$ is in int$D$ for $x_k$ sufficiently close to $x$; a contradiction.

We depict Case 2 in Figure 2 and show that it is also not possible. Note that the incenter of $\Delta(B, Y, C)$ belongs to circle $D$ iff the center $d$ lies closer to the incenter of $\Delta(B, Y, C)$ than to $Y$. With reference to Lemma 2.1, we have that the latter is true. Thus the first half of the equivalence is also true, and it contradicts the emptiness of $D$.

**Lemma 2.1.** *Let $I$ denote the incenter of $\Delta(X, Y, Z) \subset \mathbb{E}^2$. Let $P$ be an arbitrary point on the base $XZ$. Then the distance from $P$ to the incenter $I$ is less than the distance from $P$ to the opposite vertex $Y$.*

*Proof.* It is sufficient to prove that $X$, $Z$ and $I$ all lie on the same side of the perpendicular bisector of $IY$ (Figure 3); that is, the angles $\angle YIX$ and $\angle YIZ$ are obtuse. Since the sum of the internal angles of $\Delta(X, Y, Z)$ at vertices $X$ and $Y$ is less than $\pi$, we have that the sum of the internal angles of $\Delta(X, I, Y)$ at vertices $X$ and $Y$ is less than $\frac{\pi}{2}$. Thus angle $\angle YIX$ is greater than $\frac{\pi}{2}$. $\angle YIZ > \frac{\pi}{2}$ may be proved in a similar way. □

## 3. Δ is a spherical triangle

In this section, $\Delta$ denotes a spherical triangle on the surface of a sphere of unit radius. The incircle of $\Delta$ is by definition the largest area circle contained in $\Delta$. This circle meets at least two of the sides, but as Figure 4 shows, it does not necessarily meet all three sides.

We say that a triangle $\Delta$ is a $\frac{\pi}{2}$*-triangle* if each of its sides is of length at most $\frac{\pi}{2}$. Notice that the incircle of such a triangle meets each of the three sides.

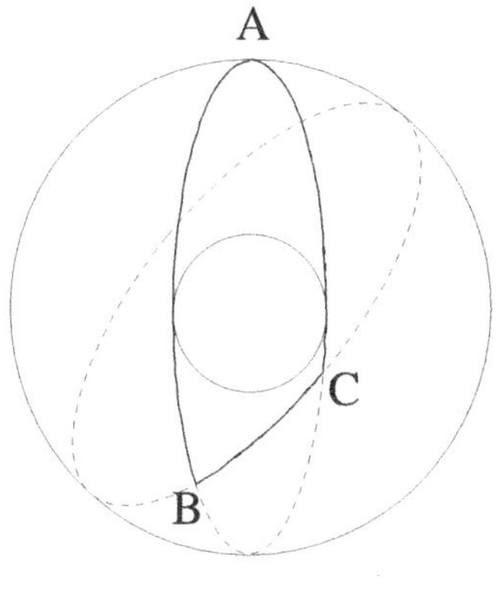


**Figure 4.**

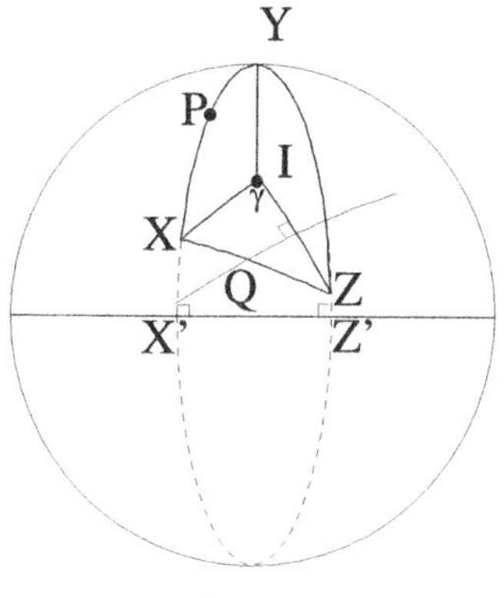


**Figure 5.**

Note that, with the exception of the proof of Lemma 2.1, the arguments used in the Euclidean case translate directly to spherical $\frac{\pi}{2}$-triangles. We prove the spherical analogue of this lemma next.

**Lemma 3.1.** *Let $\Delta(X, Y, Z) \subset \mathcal{S}^2$ be a $\frac{\pi}{2}$-triangle and denote its incenter by $I$. Let $P$ be an arbitrary point on the base $XY$. Then the distance from $P$ to the incenter $I$ is less than the distance from $P$ to the opposite vertex $Z$.*

*Proof.* It is sufficient to prove that $X$, $Y$ and $I$ lie on the same side of the perpendicular bisector of $IZ$; that is, angles $\angle XIZ$ and $\angle YIZ$ are obtuse. In view of the symmetry, it is sufficient to verify this for, say, $\angle XIZ$.

If $Y$ is considered as north pole then $\Delta(X, Y, Z)$ is contained in the upper hemisphere of $\mathcal{S}^2$, since the sides of $\Delta(X, Y, Z)$ are of length at most $\frac{\pi}{2}$ (see Figure 5). Let $X'$ and $Z'$ be points of the horizontal great circle, such that the spherical segment $YX'$ contains $X$ and the spherical segment $YZ'$ contains $Z$.

Let $Q$ denote the spherical quadrangle $XX'Z'Z$, and recall that the sum of the internal angles of $Q$ is greater than $2\pi$. From this, it follows that sum of the angles of $\Delta(X, Z, Y)$ at the vertices $X$ and $Z$ is greater than $\pi$. Thus $\gamma = \angle XIZ > \frac{\pi}{2}$. □

We have now shown Theorem 1.1 for $\frac{\pi}{2}$-triangles. Obviously Theorem 1.1 holds for all triangles which are unions of $\frac{\pi}{2}$-triangles. The fact is that one can not guarantee the existence of such triangulations for an arbitrary triangle $\Delta$. But we can show – and it is equally sufficient for Theorem 1.1 – that $\Delta$ is the limit of a sequence of closed regions, each of which is triangulated into $\frac{\pi}{2}$-triangles. What we do is very simple and natural. Let $I$ be the incenter of the triangle $\Delta(X, Y, Z)$. First we define recursively the sequence $\{X_n\}$ associated with the side $YZ$, so that $X_0$ =I and $X_{i+1}$ is the incenter of the $\Delta(YX_iZ)$. Next we define the sequences $\{Y_n\}$, $\{Z_n\}$ associated with the sides $ZX$ and $XY$ in a similar way. Finally starting with no triangles, recursively with respect to the index $n$, define a sequence of triangulations, where at each stage we add the following group of six triangles: $\Delta(X, Z_{i+1}, Z_i)$, $\Delta(X, Y_{i+1}, Y_i)$, $\Delta(Y, Z_{i+1}, Z_i)$, $\Delta(Y, X_{i+1}, X_i)$, $\Delta(Z, X_{i+1}, X_i)$ and $\Delta(Z, Y_{i+1}, Y_i)$. Obviously the limit of these triangulations is $\Delta$.

**Lemma 3.2.** *The distance between the incenter of a $\Delta$ in $\mathcal{S}^2$ and any of its vertices is at most $\frac{\pi}{2}$.*

*Proof.* Assume first that the incircle of a $\Delta$ is touching all three sides of it. Maximality of the incircle means that, within the wedge determined by any two of the sides, its center lies closer to that tip of the wedge which is the vertex of the triangle. Thus it remains to show Lemma 3.2 only for a $\Delta(A, B, C)$ whose incircle meets $CA$ and $CB$ and is disjoint from $AB$ (Figure 6).

If C is regarded as the pole then we note that the incenter $I$ lies necessarily on the equator. In view of the obvious symmetry, it is sufficient to show that $|IA| \leq \frac{\pi}{2}$. Let $A'$ be the point where the equator intersects the side $AC$. Let $M \in AC$ be a point such that $|AM| = \frac{\pi}{2}$. Note that the radius of the incircle, as the radius of any circle on the sphere, is smaller than $\frac{\pi}{2}$. Thus there is a point $A''$ on the equator with respect to pole $C$, so that $|AA''| = \frac{\pi}{2}$ and $I \in A'A''$. Since the great circular arcs $AM$ and $A'A''$ are

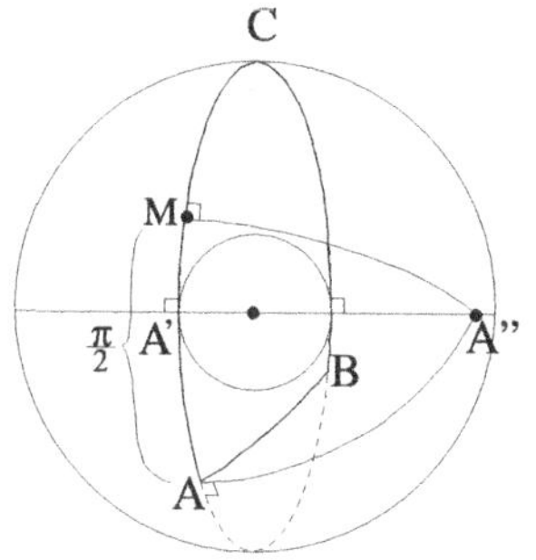


**Figure 6.**

orthogonal, the points $A$, $M$ and $A''$ are vertices of an octant. Since this octant contains $I$, and any spherical distance in an octant is at most $\frac{\pi}{2}$, we obtain that $|IA| \leq \frac{\pi}{2}$. □

## 4. Δ is a hyperbolic triangle

It is known that the incircle of a triangle in the hyperbolic plane is unique and it meets all three of the sides. It is best to use the Poincaré model to handle Cases 1 and 2. In Case 1, we may assume that the incenter coincides with the center $O$ of the model. It is then easy to check that the argument used for the Euclidean triangles is applicable in Case 1.

We depict Case 2 in Figure 7. As in the preceding section, we attempt to use the Euclidean argument. It turns out that only the hyperbolic analogue of Lemma 2.1 requires a different proof.

**Lemma 4.1.** *Let $I$ denote the incenter of $\triangle(X, Y, Z) \in \mathcal{H}^2$. Let $P$ be an arbitrary point on the base $XY$. Then the distance from $P$ to the incenter $I$ is less than the distance from $P$ to the opposite vertex $Z$.*

*Proof.* We may assume that the incenter coincides with the center $O$ of the Poincare model (Figure 8). It is known that the hyperbolic lines through $O$ are Euclidean lines,

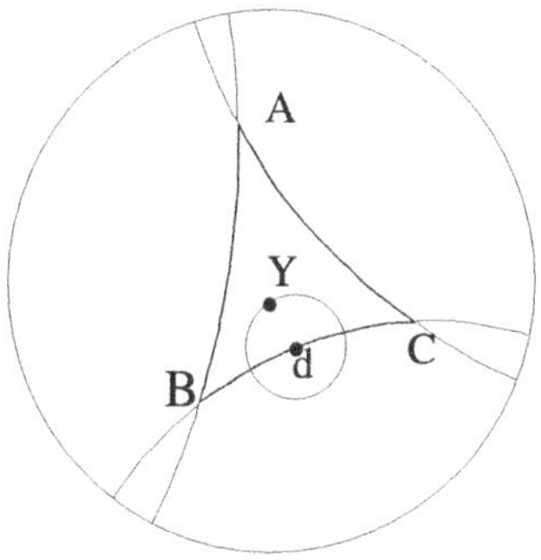


**Figure 7.**

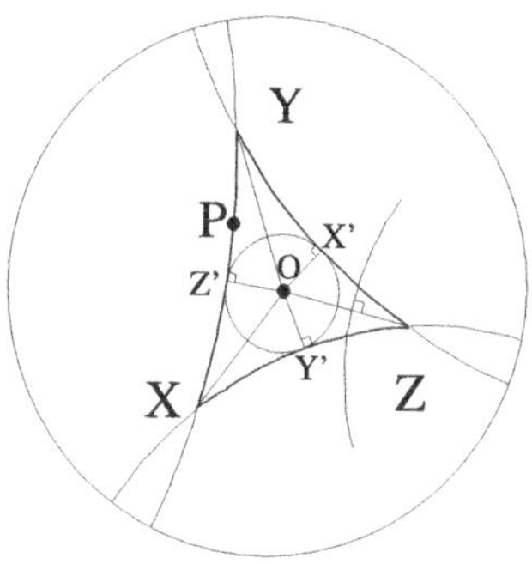


**Figure 8.**

and that the angle measure between such lines coincides with the Euclidean angle measure. It is sufficient to prove that $X$, $Y$ and $O$ lie on the same side of the perpendicular bisector of $OZ$. Since half of the internal angle of $\triangle(X, Y, Z)$ at $Z$ is less than $\frac{\pi}{2}$, it is sufficient to verify that the angles $\angle XOZ$ and $\angle YOZ$ are obtuse.

In view of the obvious symmetry, it remains to show only that $\angle YOZ \geq \frac{\pi}{2}$. We consider the perpendicular lines from $O$ to the sides, and label the feet of these perpendiculars by $X'$, $Y'$ and $Z'$. Since the sum of the internal angles of the quadrilateral $Q = Z'OY'X$ is less than $2\pi$ and since $Q$ has two right angles, we obtain that the internal angle of $Q$ at $O$ is less than $\pi$. Thus the angle completing this to $2\pi$ is greater than $\pi$, and the latter is twice the angle $\angle YOZ$. □

## References

[1] G. Ambrus and A. Bezdek, On iterative processes generating dense point sets, *Periodica Math. Hung.*, **1–2** (2006) no. 53, 27–44

[2] G. Ambrus and A. Bezdek, The solution of the three dimensional incenter problem, to appear in *Periodica Math. Hung*

[3] M. Iorio, D. Ismailescu, R. Radoičić and M. Silva, On point sets containig their triangle centers, *Computational Geometry*, **27** (2004) 257–267

Proc. Int. Conf. – Number Theory and Discrete Geometry No. 6, 2008, pp. 49–118.

# Isoperimetric Networks with Five Regions in the Euclidean Plane

Michael N. Bleicher and J. Marshall Osborn

Mathematics Department, 480 Lincoln Drive, University of Wisconsin, Madison, Wisconsin 53706
e-mail: bleicher@math.wisc.edu, osborn@math.wisc.edu

**Abstract.** In this paper we consider the following problem: Given five positive numbers, $a_i$, $i = 1, 2, 3, 4, 5$, construct a finite network which divides the plane into disjoint regions five of which have area exactly $a_i$ in such a way that the total length of the network is minimized. A region may consist of a finite number of distinct bounded components of the complement of the network. (Precise definitions are given in Section 1.) It is an open question as to whether a region must consist of one connected component cf [16]. It is also an open question whether or not there are bounded components which do not belong to any of the five regions (dead spaces) in an optimal network.

We show in this paper that for an optimal network of five regions, if each of the regions is connected, then are exactly 5 bounded components of the complement of the network; i.e., there are no dead spaces. In the process we give a much simpler proof for the cases of fewer than five regions [4,13,16,23].

*Dedicated to Professor R. P. Bambah on his eightieth birthday*

## 1. Introduction

**Definition 1.** A *network* consists of a finite number of closed rectifiable arcs in the plane, each pair of which have a finite intersection. The intersection points and end points of the arcs are called *vertices*. A part of an arc joining two vertices and containing no other vertices is called an *edge*.

**Definition 2.** The *components* of the network are the components of the complement of the network.

**Definition 3.** Let $\mathcal{P}$ be a partition of the components of a network. Exactly one element of $\mathcal{P}$, say $P_0$, contains the unbounded component. The bounded components in $P_0$ are called the *dead spaces* of the network. Each other element $P_i$ of the partition determines a (bounded, open) *region*, $R_i$, namely, the union of the set of components in it.

**Definition 4.** Given a set of $n$ positive parameters $\mathcal{A} = \{a_i : i = 1, 2, \ldots, n\}$, a network is said to *satisfy these parameters* if there exists a partition $\mathcal{P} = \{P_0, P_1, \ldots, P_n\}$ for which the area of $R_i$ is $a_i$ for positive $i$.

[1]This material was presented in preliminary form at the Joint Auburn-Budapest International Geometry Conference held at Auburn, Alabama March 21–26, 2000.

**Definition 5.** A component belonging to a region with assigned area, $a_i, i > 0$, is called an *encumbered* component. A component belonging to $P_0$ is called a *free* component.

Thus the components are in two disjoint classes, free and encumbered components.

**Definition 6.** Two components or regions are called *separated* from one another if and only if their closures have no common points.

**Definition 7.** A given network is said to be *optimal* for a set of parameters, $\mathcal{A}$, if among all networks satisfying the parameters, none have lower total length than the given network.

The existence of a solution was shown by F. Morgan [16] based on the techniques of F. J. Almgren [1]. He also proved, by different methods, the results stated in Theorem 3, first proved in [2].

**Definition 8.** Two components are called *separated* if their closures are disjoint.

**Notation.**

1. $D_0$ will denote the (unique) infinite component.
2. $D_i$, $0 < i \leq k$, will denote the dead spaces, where $k$ is the number of bounded components in $P_0 = \{D_0, D_1, \ldots, D_k\}$.
3. An encumbered component will be denoted by $E_i$ to emphasize it is a bounded component of some region $R_i$, where $i$ is positive.
4. $B_i$ will denote a bounded component.
5. $C_i$ will denote an arbitrary component.
6. If $C$ is any component of a network we use $e(C)$ to denote the *number of edges* of that component.
7. The number of dead edges of the component $E$ or $E_i$ is denoted by $d(E)$ or $d(E_i)$, respectively.

In the early history of this problem [2,3,5–12,14,17,18,21,22], the way the problem was defined did not allow for dead spaces or disconnected regions. Frank Morgan was the first to point out the alternate version of the problem which allows disconnected regions and dead spaces as possibilities. It is only by relaxing the conditions for connectivity that an existence proof is possible. This will be explained more fully below in the section on the history of the problem.

The following two conjectures were first formulated by Frank Morgan [16].

*Conjecture 1.* In an optimal network, each region consists of one component.

*Conjecture 2.* If a network is optimal for a given set of parameters, then there are no dead spaces.

Little progress has been made on these conjectures.

A more tractable conjecture is the following

*Conjecture 3.* If a network is optimal for a given set of parameters and has connected regions, then there are no dead spaces.

We note that an optimal network for any particular set of parameters must also be optimal for that set of parameters which has a separate positive number with the area of each bounded encumbered component. As a consequence every optimal network is optimal with connected regions for some set of parameters. These observations lead to

**Theorem 1.** *If every network which is optimal and has connected regions has no dead spaces, then every network which is optimal has no dead spaces.*

All three conjectures are proven for $n = 2, 3$, [2,13,15,16,19,23]. In this paper we present a proof of Conjecture 3 for the case $n = 5$ which gives simpler proofs for smaller values of $n$ and which has some hope of extending to larger values. i.e., the main theorem of this work is

**Theorem 2.** *In an optimal network of five or fewer regions, if the regions are connected, there are no dead spaces.*

## 2. An Abbreviated History of the Isoperimetric Problem

The problem of isoperimetry in the plane goes back to antiquity. The Greek mathematicians knew that a circle was the least perimeter curve to surround an given area. Zenedorus wrote a book (lost to us) on the subject in which he is alleged to have shown that the regular n-gon was the most efficient n-gon to surround a given area. The myth of Dido of Carthage shows they understood that given a natural boundary the most efficient way to surround an area with it using a given length perimeter is to draw a semi-circular arc meeting the natural boundary. The early proofs lacked rigor in the modern sense. Even after Steiner brought rigor to showing if there was a solution, it must be the circle, the existence of a solution awaited the proper definition of limit in a general context [20]. Weierstrass give the first proof, followed by Blashke's proof using compactness arguments for the space of convex sets under the Hausdorf metric [5].

Over the years the problem of dividing a given region into separate regions with given area with the minimal length was considered from several contexts. Two of these approaches were dividing the region into territories for territorial plants and animals in such a way that the length of the dividing lines was minimal and dividing fields for farmers using the least fences. There is also a free-form problem were there is no natural boundary and one wishes to fence several fields of fixed area with the least fencing material. Similarly the building construction problem of dividing a level with parallel ceiling and floor into rooms of given size. In these contexts one wishes the regions to be one connected component. An office of one hundred rooms each one square foot, say, is less useful for most purposes than a single room of 100 square feet. Similarly, farmers much prefer one large field to several smaller ones with the same total area. Thus, non-surprisingly, the early workers on this problem sought solutions with one connected region for each desired area. As more abstract methods were applied to the problem some strength was gained by the methods of Almgren as extended by F. Morgan. These techniques enabled them to prove existence, but at the price of losing control of the connectedness of the regions. This lead to dropping the requirement of connectedness and to the allowing for possible dead spaces. The two problems are both of interest and the obvious conjecture is that the solutions will be the same for a given set of area requirements with or without natural boundary.

None-the-less in view of the comments after Conjecture 3, the necessary condition on the solution of the problem with connected components in [2,3] will also hold for the more general case of solutions with dead spaces and possibly disconnected regions, since the proof shows that if the conditions of Theorem 3 don't hold, a configuration with the same component areas but shorter total length exists. Morgan obtained these results also by Geometric Measure Theoretic techniques.

## 3. Previous Results Used

For the reader's convenience we collect here the results of previous papers on related problems which we use here without proof, except for item 1, Theorem 3. References may be found in [2].

**Theorem 3.** *A network that does not satisfy all the following optimality conditions can be replaced with a shorter network with the same area divisions of its bounded components. Optimality conditions:*

1. *All the edges are segments of circles or lines; i. e., they have constant curvature.*
2. *All vertices meet exactly three edges, i. e., all vertices have degree three.*
3. *The angle between any two edges at the same vertex is* $\frac{2\pi}{3}$.
4. *The network is connected.*
5. *The sum of the signed curvature of the three edges at any vertex is zero.*

*Proof.* We give here the proof of the first statement of the theorem, since it is abbreviated in the reference. As a first step we prove the

**Lemma 1.** *If a rectifiable curve is not convex, then there are arbitrarily short segments of the curve which are not convex.*

*Proof.* We use the fact that a convex curve is characterized by the fact that as a function of the angle a support line makes with the $x$-axis, the distance of the contact point with the curve from the initial point of the curve is a monotone non-decreasing or non-increasing function depending on whether the curve is convex from above or below. Given a rectifiable non-convex curve $\Gamma$, we may add the endpoints to make it a closed set. Given a positive number, $\delta$ less than the length of the curve, we consider all open (in the relative topology) subintervals of the curve of length $\delta/2$. A finite number of these intervals will cover $\Gamma$. Since the curve is not convex the distance of the support function is not monotone, therefore either there must be one of the chosen intervals of length $\delta/2$ on which it is not monotone or there are two intervals on which it is monotone in opposite directions. In the former case we have an interval of length less than $\delta$ on which the curve is non-convex. In the latter case there must be two overlapping intervals such that the function is monotone non-increasing on one and monotone non-decreasing on the other. The function would be constant on the overlap,

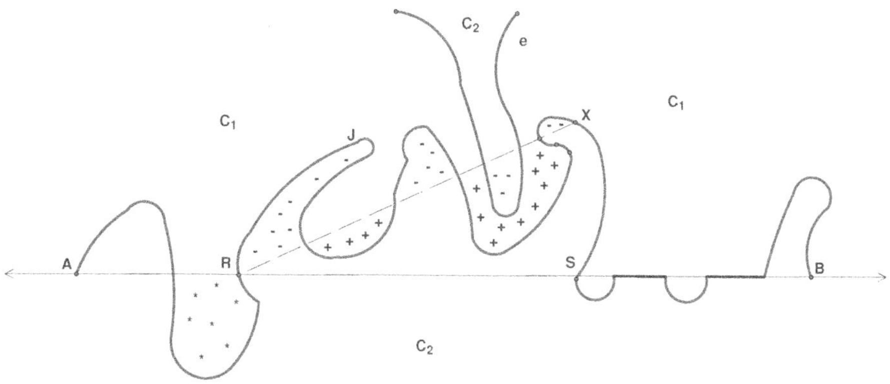


**Figure 1a.**

indicating a corner of the curve at which the curve changes from convex up to convex down or vice-versa. Taking the union of these two intervals yields an interval of length less than $\delta$ on which the curve is non-convex. □

**Lemma 2.** *In an optimal network the edges are convex curves.*

*Proof.* Let $e$ be an edge. Assume it is not convex. Then there is a closed interval, $I$, of the edge which excludes the endpoints of the edge on which the edge is non-convex. There is a minimal distance $\delta$ from this closed interval $I$ to the closed set on the network with the interior of $e$ removed. Choose an subinterval, $J$, of $I$ such that the length of $J$ is less than $\delta$ and $J$ is non-convex. Then there are two points on $J$, say $A$ and $B$, such that the chord $\overline{AB}$ has part of the edge above and part of the edge below the chord. We envision the figure with the chord $\overline{AB}$ horizontal. See Figure 1a. Let $C_1$ and $C_2$ be the components separated by $e$ with $C_1$ above the curve and $C_2$ below. Consider the closed subsets $V$ and $W$ of $\overline{AB}$ given by given by

$$V = \overline{AB} \cap \overline{C_1}$$

and

$$W = \overline{AB} \cap \overline{C}_2.$$

Neither of these subsets is all of $\overline{AB}$, consequently both

$$V' = \overline{AB} \setminus W \quad \text{and} \quad W' = \overline{AB} \setminus V$$

are non-empty open subsets of the line segment $\overline{AB}$. As open subsets of a line segment each of $V'$ and $W'$ can be written as a union of open intervals. Pick one interval, say $I_V$ and $I_W$, from each of $V'$ and $W'$, respectively. Since $e$ has no self intersections, these intervals are disjoint and contain no points of $e$, but the endpoints are on $e$. The curves $\Gamma_V$ and $\Gamma_W$ consisting respectively of the intervals $I_V$ and $I_W$ each united with the part of $e$ between their respective endpoints are both Jordan curves and surround domains with area $a_V$ and $a_W$, respectively. If $a_V = a_W$, we replace the part of $\Gamma_V$ and $\Gamma_W$ not

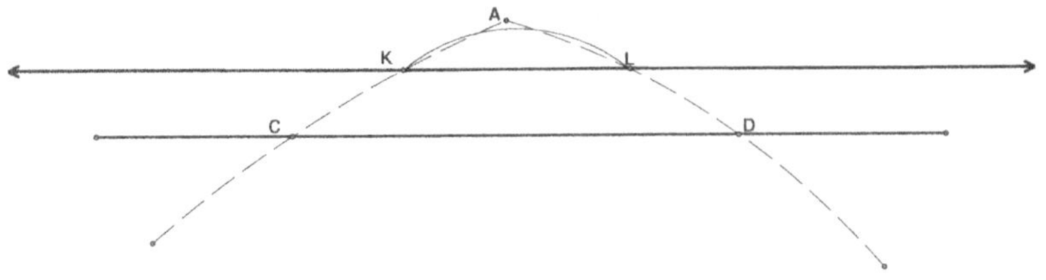


**Figure 1b.**

on $\overline{AB}$ by $I_V$ and $I_W$, which results in the same area division of $C_1$ and $C_2$, but with a shorter length network, contradicting the optimality. If $a_V \neq a_W$ we may suppose without loss of generality that $a_V < a_W$. The area of $a_V$ is filled with asterisks in the figure. Let the endpoints of $I_W$ be $R$ and $S$. Consider the variable chord, $\overline{RX}$ with one endpoint at $R$ and the other varying along the arc $\Gamma_W \cap e$ from $S$ to $R$. We note that $X$ is always above the line through $\overline{AB}$. See Figure 1a. We now redefine the edge $e$ to obtain a new edge $e_X$ as follows:

1) Replace the part $\Gamma_W$ between $R$ and $X$, denoted by $\Gamma(X)$ by the line segment $\overline{RX}$.
2) If there are intersections of $e \setminus \Gamma(X)$ with the line segment $\overline{RX}$ they will divide the segment into open intervals for which a segment of $e \setminus \Gamma(X)$ joins the endpoints of the segment. Delete these segments of $e \setminus \Gamma(X)$.

This may add some areas of $C_1$ to $C_2$, marked with plus signs and some areas of $C_2$ to $C_1$, marked with minus signs. Let $a(X)$ be the function of $X$ giving the signed area change to $C_1$ as a function of $X$. We note that the change of area from replacing $\Gamma_V$ by $I_V$ increased the area of $C_2$ by $a_V$. If we choose $X = S$ the replacement process for $\overline{RX}$ subtracts too much area from $C_2$, since $a(X) = -a_W$ and $a_V < a_W$. If we choose $X = R$ then $a(X) = 0$ and we subtract too little, but since $a(X)$ is a continuous function of $X$ there will be a choice for which $a(Y) = -a_V$. The new boundary for this choice will leave the areas of $C_1$ and $C_2$ unchanged, but shorten the network length. This is a contradiction to optimality. $\square$.

Figure 1a does not show all the combinations. For example, there may be infinitely many intervals on $\overline{RX}$ or the tongue of $C_2$ might curl back and forth across $\overline{RX}$.

We now prove the first statement of Theorem 3.

Suppose, by way of contradiction, that an optimal network has an edge $\Gamma$ which does not have constant curvature. In this case some interior point of $\Gamma$ has either no curvature or no derivative of the curvature or a non-zero derivative of the curvature. Let $A$ be such a point. Consider a closed interval, say $I$ on $\Gamma$, centered at $A$ which does not contain the endpoints of $\Gamma$. Let the endpoints of $I$ be $C$ and $D$. See Figure 1b There is a minimum positive distance from this closed interval to the closed set consisting to the closed network with the interior of $\Gamma$ removed. Let this minimal distance be $\delta$. Pick an interval, say $J$, inside $I$ centered on $A$ of length less than $\delta$. Label its endpoints $K$ and $L$ By the well known "Stick and String Problem" [17], we can replace the interval $J$ of $\Gamma$ by a circular arc with $\overline{KL}$ a chord of the circle from which the arc is taken and which encloses the same area as $\overline{KL}$ and $J$. The circular arc is shorter and yields the same area division on the two sides of $\Gamma$. Because of the convexity of $\Gamma$, the circular arc will not meet any of the curve $\Gamma \setminus J$ since the circular arc is above the chord joining the endpoints of $J$. Because the length of this circular arc is less than $\delta$, it will not touch any other edges of the network. But this produces a new network of shorter length with the same area division, a contradiction to the optimality of the network. Therefore there exist no edge of an optimal network which does not have constant curvature. □

**Lemma 3.** *Consider an optimal network and a directed simple closed curve, $\Gamma$, meeting the network finitely often and containing no vertices of the network. Traversing $\Gamma$, we assign* signed curvature *to each edge crossed by giving it the sign from the component just before it is crossed. The sum of the signed curvatures of all the edges crossed by $\Gamma$, counting multiplicities, is zero.*

**Lemma 4.** *In an optimal network free spaces are separated. That is, for $0 \leq i < j \leq k$ we have $\overline{D}_i \cap \overline{D}_j = \emptyset$.*

**Lemma 5.** *In an optimal network two distinct components of the same region are separated.*

**Lemma 6.** *In an optimal network two distinct components can not have two edges in common.*

**Notation.**

1. Since edges have constant curvature, we denote the curvature of the edge $e$ or $e_i$ by $\kappa$ or $\kappa_i$, respectively.
2. The total curvature of a circular arc is the central angle it subtends.
3. The *signed total curvature* of an edge $e$ or $e_i$ denoted by $\tau$ or $\tau_i$, respectively, is the total curvature with a plus or minus sign depending on whether it is viewed from the side on which the center of the circle lies (+) or the opposite side (−). Thus it depends on both the edge and the component from which it is viewed. The total curvature of a line segment is zero.

**Lemma 7.** *At any node at which no edge has zero curvature, one of the three components meeting there has both edges of positive curvature, one has both negative, and the third has one positive and one negative.*

Let $v(C)$ be the number of vertices on the perimeter of $C$. Note: $v(C) = e(C)$ unless $e(C) = 1$, in which case $v(C) = 0$.

**Lemma 8.** *Let $n$ be the number of encumbered regions of an optimal network of connected regions. For an optimal network the following hold:*

1. *If $\tau_i$ is the total curvature of the $i^{th}$ edge of a bounded component, $B$, oriented from inside that component, then*

$$\sum_{i=1}^{v(B)} \tau_i = 2\pi - \frac{v(B)\pi}{3}.$$

2. *If $\tau_i$ is the total curvature of the $i^{th}$ edge of the unbounded component, oriented from the inside of the complement of the unbounded component, then*

$$\sum_{i=1}^{v(D_0)} \tau_i = 2\pi + \frac{v(D_0)\pi}{3}.$$

3. $e(D_0) = 1$ *if and only if there is a bounded component, $B$ for which $e(B) = 1$ if and only if $n = 1$.*
4. $e(D_0) = 2$ *if and only if there is a bounded component, $B$ for which $e(B) = 2$ if and only if $n = 2$.*
5. $e(D_0) \geq 3$ *if and only if for every bounded component, $B$, $e(B) \geq 3$ if and only if $n \geq 3$.*

In Item 1. and 2. of the preceding lemma we shall normally use $e(C)$ in place of $v(C)$, since we are concerned with connected networks of more than one bounded component.

**Lemma 9.** *If $D$ is a free region in a network with $n$ encumbered regions, all of which are connected, then $e(D) \leq n$.*

**Lemma 10.** *In an optimal network of $n$ regions, all of which are connected, the number of edges of a bounded region is at most $2(n-1)$; i.e., $e(E) \leq 2(n-1)$.*

The following lemma and proof are included in this section, even though it is not found in the references, since it is with them in spirit.

**Lemma 11.** *In an optimal network all the dead edges of an encumbered component have the same curvature.*

*Proof.* If two dead edges had differing curvatures we cut each dead edge with sufficiently small equal length chords and interchange the arcs these chords subtend. This does not change the area of the encumbered region or the length of the network. But this network can not be optimal, since the two new edges do not have constant curvature. Replacing sufficiently short parts these edges by curves of constant curvature which surround the same area yield a network of less length satisfying the parameters. Thus the original network could not have been optimal. □

In view of Theorem 1, we concentrate or efforts on optimal networks with connected regions. Henceforth when we say **optimal network**, it is to be understood that we mean **with connected regions.**

## 4. The Pathways of a Network

We now superimpose a directed graph onto the network as follows. Pick a point in each component, say the centroid for the bounded components and any point for the unbounded one. In each component join the chosen points by non-intersecting arcs to the center of each of its edges for which the curvature is non-zero.

**Definition 9.** The *vertices* of the directed graph are the points chosen in the preceding paragraph.

**Definition 10.** An *edge* of the directed graph is the union of the two arcs joining the vertices of two adjacent components to the midpoint of their common edge, if the common edge is not a line segment.

**Definition 11.** The *direction of a graph edge* is that direction in which the arc crosses from the component in which the edge has positive curvature into the component in which the curve has negative curvature.

**Definition 12.** A *path of length $m$* is a sequence $\gamma_1, \gamma_2, \ldots, \gamma_m$, of distinct, adjacent edges joining the vertices in such a way that traversing them in the positive sense, the end of $\gamma_i$ is the same point as the beginning of $\gamma_{i+1}$.

**Lemma 12.** *In an optimal network, no path can return to the same vertex; i.e., there are no circuits.*

*Proof.* If a circuit exists, then consider a shortest one. For positive $\epsilon$ sufficiently small, at each edge crossed, replace the circular arc by a small line segment which reduces the area of the beginning component by $\epsilon$ and increase the area of the ending component by $\epsilon$. Since in a minimal circuit, each component is both the beginning and end of exactly one arc, the area of each component is unchanged, but the length of the network is decreased, a contradiction to optimality. □

**Definition 13.** A *maximal path* is a path which is not part of any longer path.

Since there are only finitely many edges and no circuits, all paths are finite in length and the length is bounded by the number of vertices. Thus maximal paths exist.

**Lemma 13.** *In an optimal network, a maximal path can begin only in a region with no edges of negative curvature, and can end only in a region with no edges of positive curvature.*

*Proof.* Otherwise the path could be extended, contradicting the maximality. □

**Definition 14.** A component with a positive curvature edge and no edge with negative curvature is called a *source*. A component with a negative curvature edges and no positive curvature edges is called a *sink*.

In this terminology, we see that all maximal paths begin in a source and end in a sink.

**Lemma 14.** *No path can go through more than one free components.*

*Proof.* The two free components can not be adjacent, so there are encumbered components in between. We do the same trick as in the proof of Lemma 12. This will change the areas of the two free regions, but that does not matter; the parameters will still be satisfied. Hence the optimality will be contradicted. □

**Theorem 4.** *In an optimal network of at most five connected regions, a bounded component can not be a sink.*

This theorem is proved by a series of lemmata. Showing first that no dead space can be a sink, Lemma 17 , then that no encumbered component can be a sink, Theorem 5. The latter takes several cases.

**Lemma 15.** *A bounded sink has at least 7 edges.*

*Proof.* Since none of the edges of a sink have positive curvature and at least one edge has negative curvature, the total curvature sum is negative. On the other hand by Lemma 8, we have

$$\sum_{i=1}^{m} \tau_i = 2\pi - \frac{m\pi}{3},$$

where m is the number of edges of the bounded sink $B$. Since the left-hand side is negative, $m \geq 7$. □

**Note.** For $n \leq 4$, Lemma 10 and Lemma 9 prove the previously known results on the non-existence of bounded dead spaces. This follows since $D_0$ will be the only sink. If a dead space exists, it would have at least one edge of positive curvature, so a path could begin there and proceed until it ended at the only sink $D_0$, violating Lemma 14.

**Lemma 16.** *In an optimal network of at most 5 encumbered regions, $e(D) \leq 5$, for any free component D.*

*Proof.* This is the special case of Lemma 9 for $n = 5$. □

These last two lemmata yield

**Lemma 17.** *No dead space can be a sink.*

We next prove

**Theorem 5.** *No encumbered component can be a sink.*

The proof of this theorem will take several lemmata. We begin with

**Lemma 18.** *A bounded sink is an encumbered region of seven or eight edges.*

*Proof.* By Lemma 6 a given encumbered component can meet each of the other encumbered components in at most one edge. By Lemma 4 the contacts with free spaces must be separated by encumbered components, hence there can be at most four contacts with dead components. Thus the total number of edges is at most 8. This combined with Lemma 15 yields this lemma. □

We next show that such a sink is impossible.

We use the following labeling scheme for the remainder of the proof. Let $E$ denote the hypothetical sink. If $e(D) = 7$ we name the faces around $E$ going clockwise by $E_1, D_1, E_2, D_2, E_3, D_3, E_4$. See Figures 1. If $E(d) = 8$, we name the faces around $E$ by $E_1, D_1, E_2, D_2, E_3, D_3, E_4, D_4$. See Figure 2. We use the convention that a thick arc denotes an edge for which the sign of the curvature is undetermined, at least by the presented arguments. A dashed line is one for which the sign of the curvature is determined by the arguments presented or apparent. Thin arcs are the edges of the hypothetical bounded sink, $E$. We occasionally use a thin line for a separating path which may be more than one edge. We sometimes slightly abuse the notation by saying $D_0 = D_i$ to mean that $D_0$ meets $E$ where $D_i$ was, and $D_i$ does not exist.

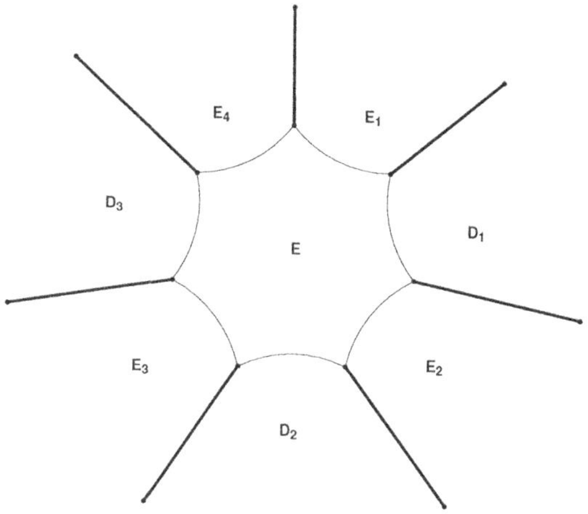


**Figure 2a.**

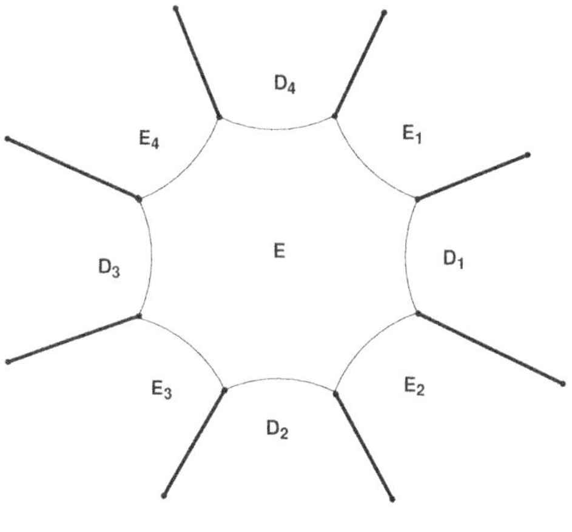


**Figure 2b.**

**Lemma 19.** *If E is the sink, then one of the dead spaces meeting it must be three-edged, except in the following two situations and their left-right mirror images:*

a) $e(E) = 7$, $D_0 = D_1$ *and the edge separating* $E_1$ *and* $E_2$ *has one end on* $D_0$ *and the other on* $D_2$. *Figure 5d shows the only way this configuration fails to have a three-sided dead space.*
b) $e(E) = 7$, $D_0 = D_2$ *and the edge separating* $E_2$ *and* $E_3$ *has one end on* $D_0$ *and the other on* $D_1$. *Figure 6b shows the only exceptional configuration.*

*Proof.* When $e(E) = 8$, if $D_0$ meets $E$, by symmetry we may suppose $D_0 = D_4$. If $D_2$ has 3 edges, we are done. If $D_2$ is not three-sided, it will have either 4 or 5 edges. It follows that at least one of $E_1$ or $E_4$ must meet it. Without loss of generality, suppose it is $E_1$. In this case, $D_1$ is separated from $E_4$ and $E_3$ so $D_1$ must be 3-sided. See Figures 3a for one possible way $E_1$ can meet $D_2$; the thin line separating $E_1$ and $E_4$,

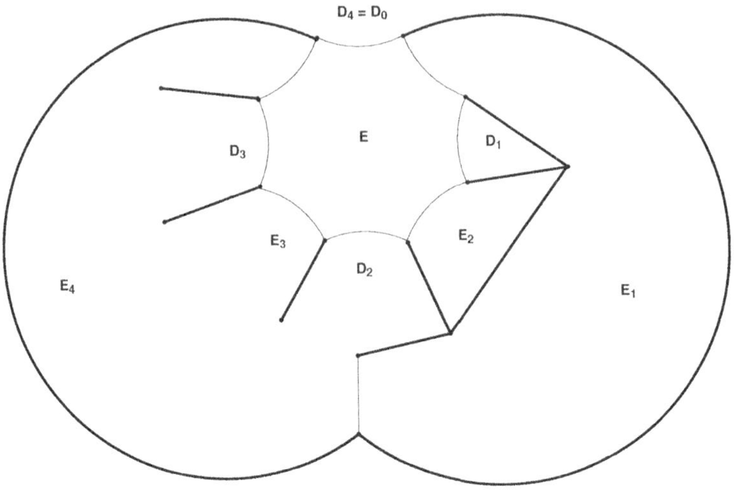


**Figure 3a.**

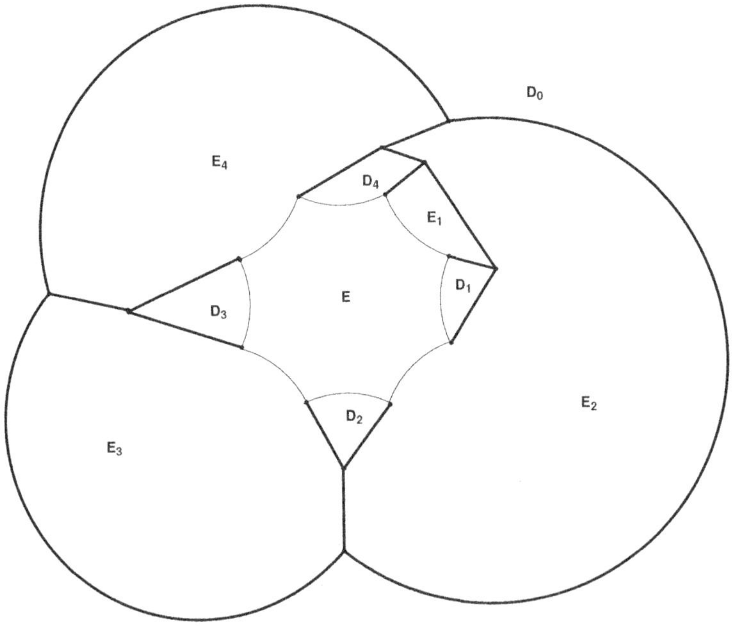


**Figure 3b.**

in this particular instance, indicates that it might represent several edges, if $E_1$ meets more components.

If $D_0$ is separated from $E$, then at least three of the $E_i$ must meet $D_0$, some pair of these $E_i$ meeting both $E$ and $D_0$. are separated by a dead space. The dead space between these two can only meet these two $E_i$, thus this dead space is 3-sided. Figure 3b shows a configuration when $E_3$ and $E_4$ meet both $E$ and $D_0$.

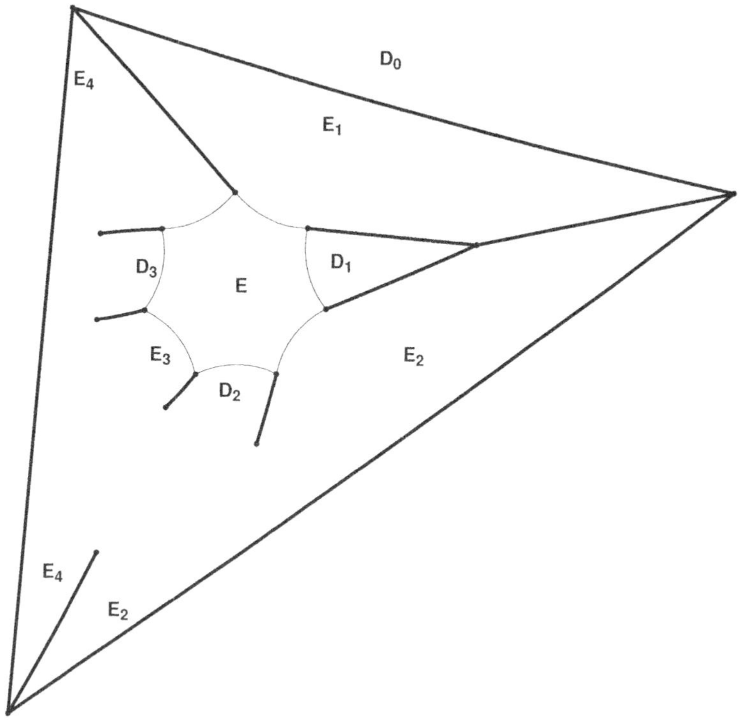


**Figure 4a.**

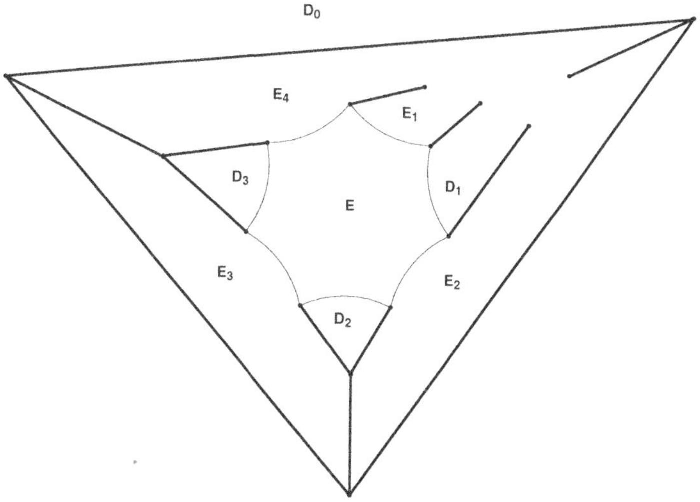


**Figure 4b.**

When $e(E) = 7$ and $D_0$ is separated from $E$, the picture is slightly different, but the argument is similar. Three of the $E_i$ meet both $E$ and $D_0$, and some pair of these will surround a dead space; The components meeting $D_0$ might include the one or both $E_1$ and $E_4$. Allowing for symmetries, Figures 4a and 4b show the two possibilities. There

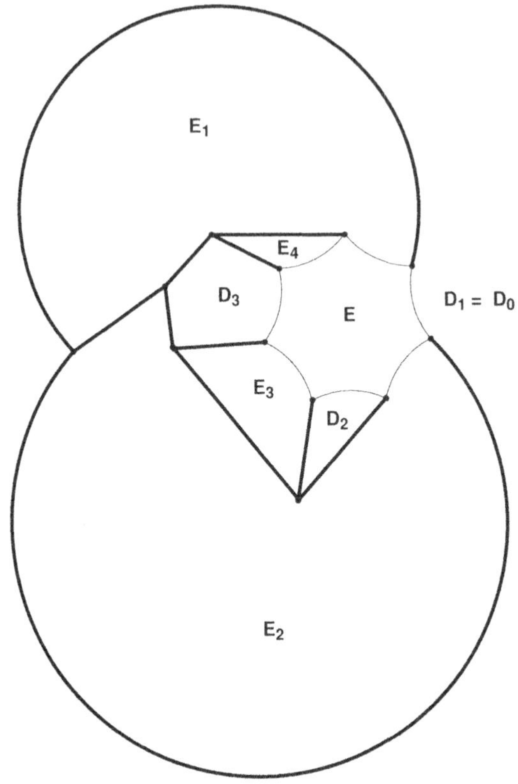


**Figure 5a.**

is a three-edges dead space in each. The open cells indicate that the edges of these cells are not completely determined, but the three-edged dead space is already determined.

If $D_0$ meets $E$ and $e(D_0) \geq 4$ then again two $E_i$ with a dead space between meet both $E$ and $D_0$. The surrounded dead space must then be three-sided. Finally, we consider when $e(E) = 7$ and $E$ meets $D_0$ with $E(D_0) = 3$. In this case we have either $D_0 = D_1$ or $D_0 = D_3$ or $D_0 = D_2$. The first two options are symmetric, we consider only the first and third. When $D_0 = D_1$, we consider the possible endpoint of the edge,say $e_{1,2}$, beginning on $D_0$ and separating $E_1$ from $E_2$. If $e_{1,2}$ ends on $D_3$, Figure 5a, we see that $D_2$ is separated from both $E_1$ and $E_4$, so can meet only three components. If the edge $e_{1,2}$ ends on $E_3$, Figure 5b, then $E_4$ and $E_1$ are again separated from $D_2$, which thus has only three sides. (There could be another three-sided dead space where $E_1$, $E_2$, and $E_3$ meet, but it does not effect this argument.) If $e_{1,2}$ meets $E_4$ then $E_2$ must meet $D_3$ or $D_0$ would have three edges, but then $D_2$ is again separated from $E_4$ and $E_1$, Figure 5c. (Again, there might be a dead space where $E_1$, $E_2$, and $E_4$ meet.) If $e_{1,2}$ meets $D_2$, then there may be no three-sided dead space. Figure 5d shows the only way this can happen, except for symmetries when $D_0 = D_3$.

If $D_0 = D_2$, the situation varies with the end point of the edge $e_{2,3}$ which separates $E_2$ from $E_3$. If $e_{2,3}$ end on $E_1$ (which is symmetric with $E_4$) then $D_1$ ($D_3$) has three edges, Figure 6a shows this possibility when $e(D_3) = 4$. If $e_{2,3}$ ends on $D_1$ (which is

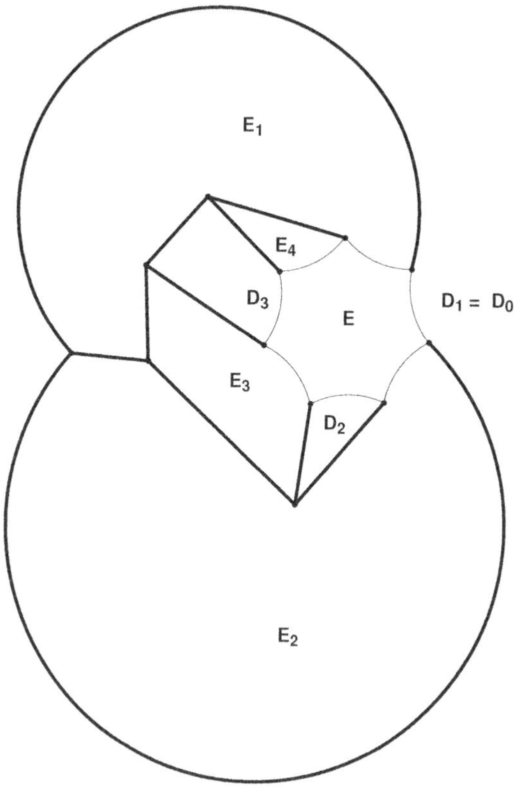


**Figure 5b.**

symmetric with $D_3$), Figure 6b shows the only way there might not be a three-sided dead space. □

**Lemma 20.** *All the edges separating the sink from a dead space have negative curvature from the perspective of the sink.*

*Proof.* Since no edge of a sink can have positive curvature, if the edge does not have negative curvature, it is a line segment. Suppose, by way of contradiction, one edge with a dead space has zero curvature, Since all edges of dead spaces meeting an encumbered region have the same curvature by Lemma 11, they all have zero curvature and hence zero total curvature. Consider a bounded, three-sided dead space meeting $E$, which exists by Lemma 19, except for the two special cases discussed in the next paragraph. Since the sum of the curvatures of the edges meeting at a vertex is zero, the other two edges of this three-sided dead space are of non-positive curvature from the perspective of the dead space. But the sum of the total curvatures of edges of a three-sided region must be $\pi$ by Lemma 8, a contradiction.

If there is no three-sided dead space meeting $E$ then Figures 5d and 6b show the possibilities (up to symmetries). In Figure 5d, we first note that the edges of $D_0$ have the sum of total curvature of its edges equal to $3\pi$ by Lemma 8, which means the edges of $D_0$ with $E_1$ and $E_2$ have positive curvature from the perspectives of $E_1$ and $E_2$,

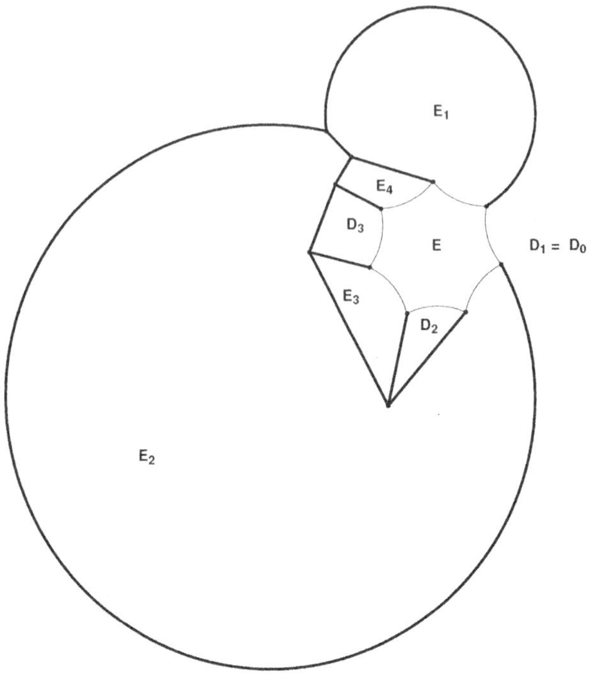


**Figure 5c.**

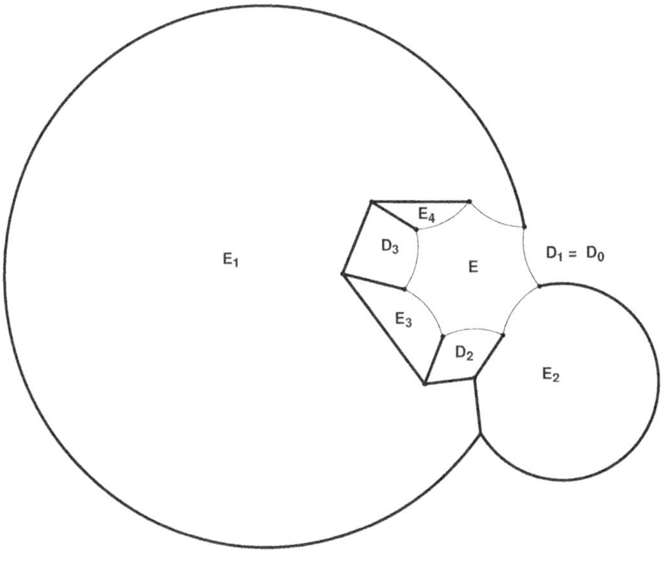


**Figure 5d.**

respectively. But they have the same curvature with every dead space they meet, so they have positive curvature with $D_2$. So those two edges have negative curvature from the perspective of $D_2$. Furthermore if the edge separating $D_2$ from $E$ had zero curvature,

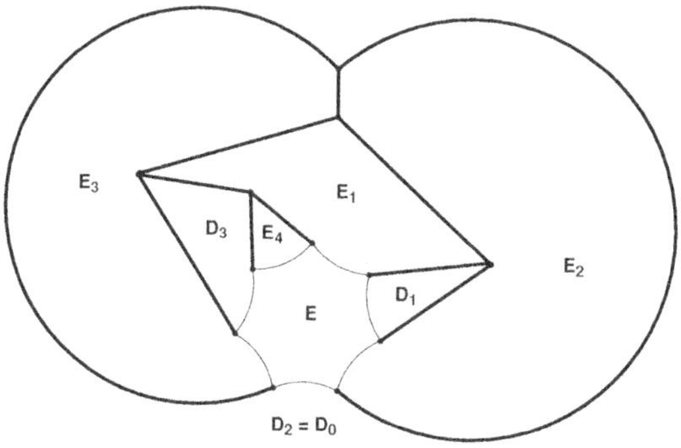


**Figure 6a.**

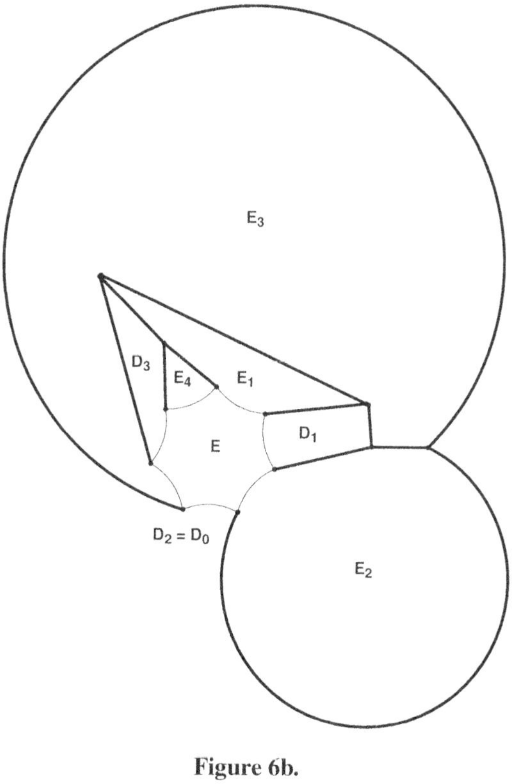


**Figure 6b.**

examining the vertex where its edge with $E_3$ meets $E$, we deduce from Theorem 3 (5) that the edge has non-positive curvature from the perspective of $D_2$. Thus no edge of $D_2$ would have positive curvature, violating Lemma 8. Similar arguments apply to $D_1$ in Figure 6b. It is thus not possible for the edges of $E$ with its adjoining dead spaces to have zero curvature. □

Let $E$ be a bounded sink. We consider ten cases according as $e(E)$ is seven or eight, $E$ meets $D_0$ or not, and the size of $e(D_0)$. Most of these cases will be broken into subcases.

**Lemma 21.** *If $D_0$ does not meet $E$, then $D_0$ has either 3 or 4 edges. If $D_0$ meets $E$, then $D_0$ has 3, 4, or 5 edges.*

*Proof.* If $D_0$ does not met $E$, then, since it can meet only the five encumbered component, once each and does not meet $E$, there are at most four edges. Also it may not have fewer than 3 edges. If $D_0$ meets $E$ then it may meet from two to four of the other encumbered components. □

We now consider the ten major cases as follows:

Table of Cases

| Case# | e(E) | $D_0$&$E$ | $e(D_0)$ |
|---|---|---|---|
| 7m3 | 7 | meet | 3 |
| 7m4 | 7 | meet | 4 |
| 7m5 | 7 | meet | 5 |
| 7s3 | 7 | separated | 3 |
| 7s4 | 7 | separated | 4 |
| 8m3 | 8 | meet | 3 |
| 8m4 | 8 | meet | 4 |
| 8m5 | 8 | meet | 5 |
| 8s3 | 8 | separated | 3 |
| 8s4 | 8 | separated | 4 |

We turn first to the cases when $e(E) = 7$. In these cases the sum of the total curvature of the edges of $E$ is $-\pi/3$; so every edge with $E$ has total curvature, $\tau$ satisfying $|\tau| < \pi/3$. This, with Lemma 8, implies that for a three-edged dead space meeting $E$, the sum of the total curvature of the other two edges exceeds $2\pi/3$. in particular if a second edge has non-positive curvature, the third edge has curvature which exceeds $2\pi/3$. In view of Lemma 20, for any collection of edges of $E$ which does not include all the dead edges the sum of the total curvatures of the edges in the collection is greater than $-\pi/3$.

The cases where $e(E) = 7$ are complicated by the lack of some of the symmetry of the cases where $e(E) = 8$.

In outlining the subcases of the major subcases we use the following

**Notation.**

1. If $C_i$ and $C_j$ are two components then $C_i \| C_j$ indicates they are separated.
2. The notation $\neg(C_i \| C_j)$ indicates $C_i$ and $C_j$ meet, i.e., they have a common edge.
3. If $E_i$ and $E_j$ are two encumbered regions which meet we use $e_{i,j}$ to denote the edge they share.
4. If $E_i$ and $D_j$ are an encumbered component and a free component which meet, the common edge is denoted by $e_{i,j*}$.
5. If $C_i$ and $C_j$ meet then $C_i \to C_j$ means there is an edge of the direct graph going from $C_i$ to $C_j$
6. If $C_i$ and $C_j$ meet then $\neg(C_i \to C_j)$ means there is not an edge of the direct graph going from $C_i$ to $C_j$. That is, either the edge has zero curvature or the directed graph goes the other way.

7. If $e_{i,j}$ is an edge then $e_{i,j} \Rightarrow C_k$ means that one end of the edge is on $C_k$. Similarly for an edge $e_{i,j*}$.

**Case 7m3:**

This case and Case 8m3 are the most difficult (tedious) cases. We will do this one in full detail.

We use the labeling system described in Figure 1.

In this case we have two major subcases, depending on whether $D_0 = D_1$ (which is symmetric to $D_0 = D_3$) or $D_0 = D_2$.

**Subcase 7m3a:** $D_0 = D_1$

The following is a outline of the subcases of Subcase 7m3a.

a) $D_0 = D_1$
   - i) $e_{1,2} \Rightarrow D_2$
     - A) $E_4 \| E_3$
     - B) $\neg(E_4 \| E_3)$
       - 1) $E_4 \| D_2$
       - 2) $\neg(E_4 \| D_2)$
   - ii) $e_{1,2} \Rightarrow E_3$
     - A) $e_{1,3} \Rightarrow E_4$
       - 1) $E_4 \rightarrow E_3$
       - 2) $\neg(E_4 \rightarrow E_3)$
     - B) $e_{1,3} \Rightarrow D_3$
   - iii) $e_{1,2} \Rightarrow D_3$
   - iv) $e_{1,2} \Rightarrow E_4$
     - A) $e_{2,4} \Rightarrow D_3$
     - B) $e_{2,4} \Rightarrow E_3$
       - 1) $E_3 \rightarrow E_4$
     - C) $e_{2,4} \Rightarrow D_3$
       - 1) $D_2 \rightarrow E_4$
       - 2) $\neg(D_2 \rightarrow E_4)$

The figures for this subcase will vary with the location of the endpoint of the edge $e_{1,2}$ separating $E_1$ from $E_2$. The endpoint might be on any of $D_2$, $E_3$, $D_3$, or $E_4$. See Figure 7m3a where the curvature of the edges $e_{1,0*}$ and $e_{2,0*}$ follow from Lemma 8 (2).

If the edge $e_{1,2}$ end on $D_2$, we can deduce the curvature of the edge of $D_2$ with $E_1$ and $E_2$, Figure 7m3ai. The dashed arcs have the indicated sign for their curvatures;

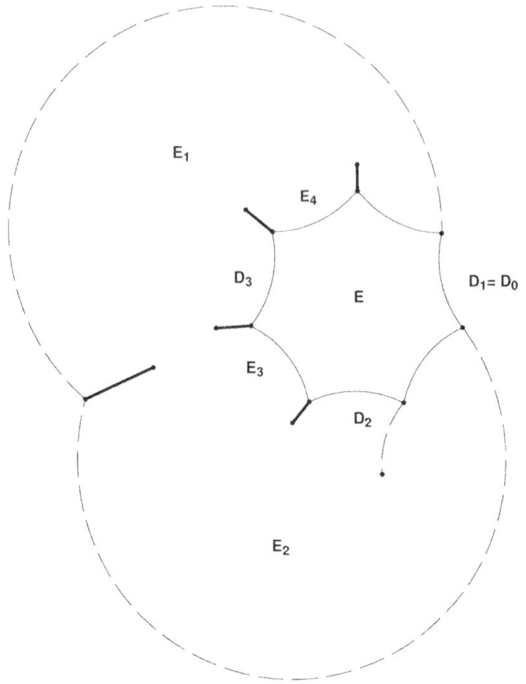


**Figure 7m3a.**

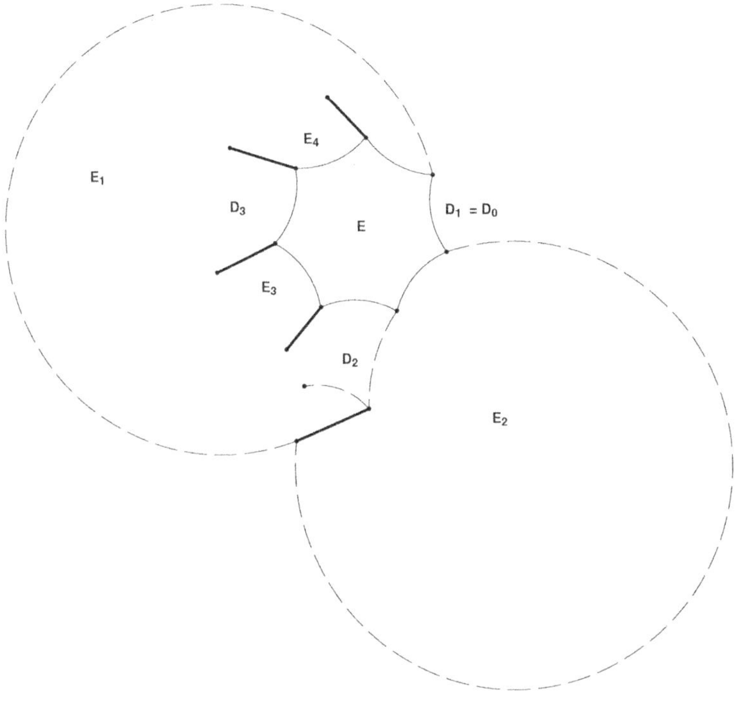


**Figure 7m3ai.**

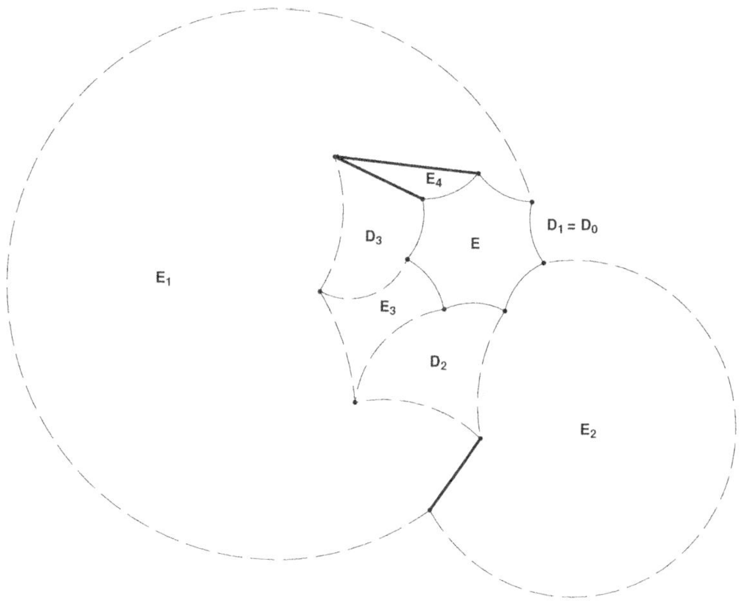


**Figure 7m3aiA.**

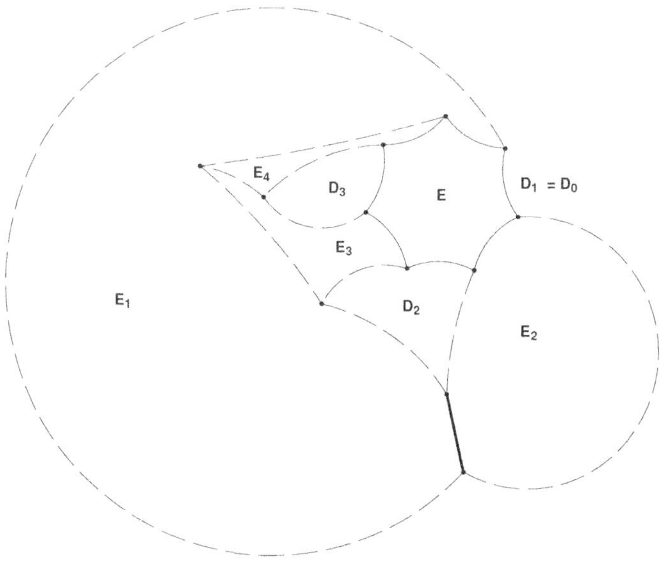


**Figure 7m3aiB1.**

the heavy lines have undetermined (at least as far as the presented argument) curvature sign. We don't know if $E_4$ meets or $E_3$.

If $E_4$ does not meet $E_3$, then $E_1$ must separate them at $D_3$, so $E_4$ also does not meet $D_2$. See Figure 7m3aiA. In this instance, $D_2$ must have an additional edge with positive curvature, namely the edge with $E_3$, which then determines the curvature of the edge between $E_3$ and $D_3$. The vertex of $D_2$ then determines the curvature of the

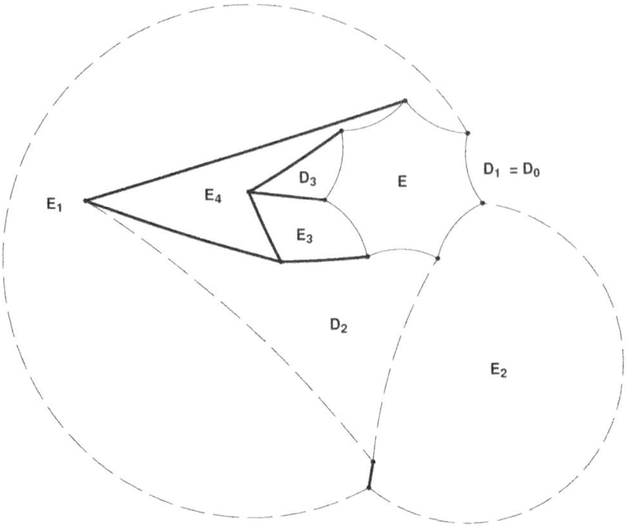


**Figure 7m3aiB2.**

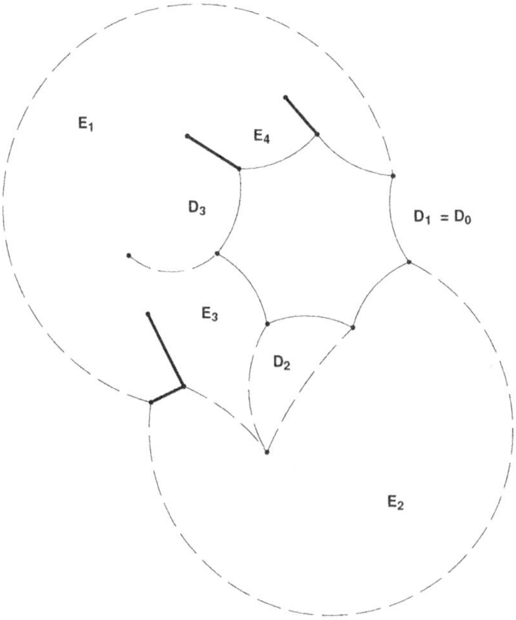


**Figure 7m3aii.**

edge separating $E_3$ from $E_1$ via Theorem 3 (5). But then the only edge of $E_3$ with positive curvature is with $E$, but the total curvature any edge of $E$ is insufficient for the total curvature sum of this four-edged component to be $4\pi/3$ which is required in an optimal network according to Lemma 8.

We now suppose $E_4$ meets $E_3$. If $E_4$ does not meet $D_2$, then its edge with $E_3$ must be positive to avoid the contradiction of the last paragraph, Figure 7m3aiB1. But then

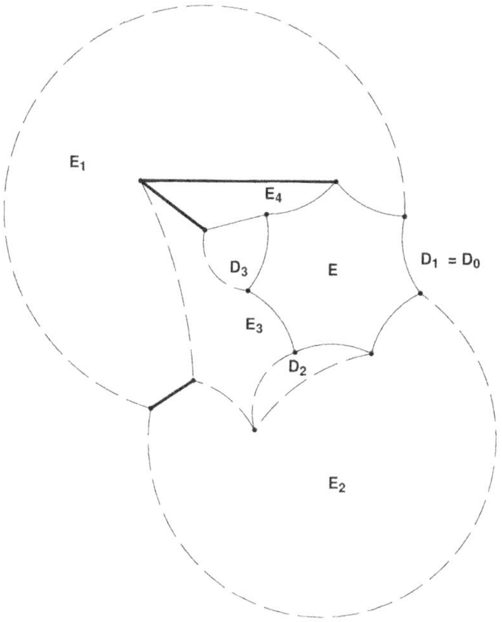


**Figure 7m3aiiA.**

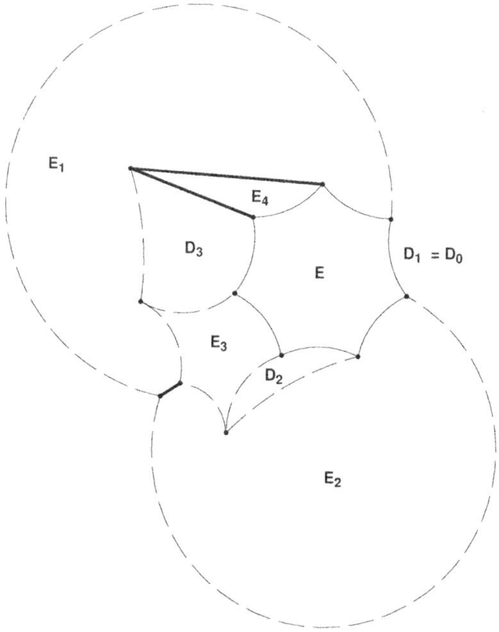


**Figure 7m3aiiB.**

the other edges of $E_4$ have the positive curvature shown in Figure 7m3aiB1. But now we have a contradiction to the sum of the total curvatures of the four-sided $E_4$. There might be an additional dead space at the conjunction of $E_1$, $E_3$, and $E_4$. If so, we include that dead space with $E_4$ to get the contradiction.

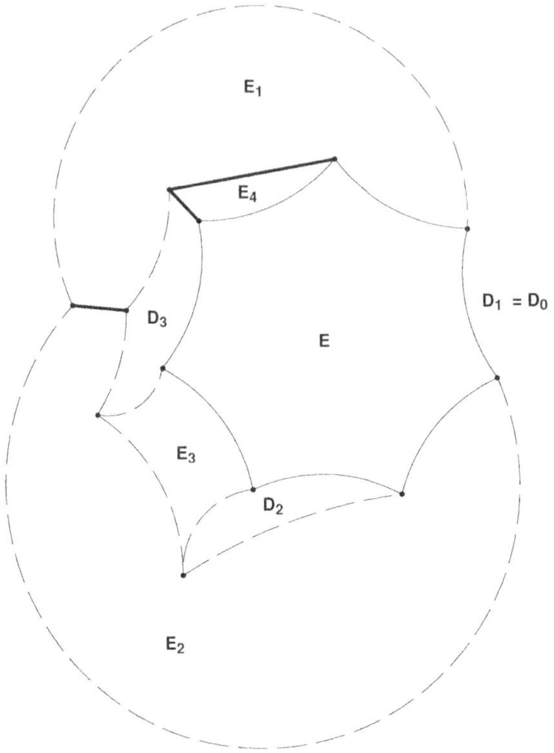


**Figure 7m3aiii.**

If $E_4$ meets $D_2$, Figure 7m3aiB2, we consider $E_1 \cup E_2$. Summing the total curvatures separately we see from Lemma 8 that the sum of the total curvature of the edges of these two components, one with four edges and the other with five edges, is $\pi$. But the positive total curvature of the edges with $D_0$ exceeds $3\pi$, The only other edge of $E_1 \cup E_2$ with possible negative curvature is the edge with $E_4$, but the total curvature of any single edge is greater than $-2\pi$. Thus the sum of the outer edges of $E_1 \cup E_2$ is greater than $\pi$, a contradiction to optimality. Thus the edge separating $E_1$ and $E_2$ can't meet $D_2$.

Consider the possibility that $e_{1,2}$ ends on $E_3$, Figure 7m3aii. We know $E_1$ and $E_3$ have a common edge. We do not know whether the other end of $e_{1,3}$ lies on$D_3$ or $E_4$. If $E_{1,3}$ ends on $E_4$, Figure 7m3aiiA, then the vertex of $D_2$ determines the sign of the curvature of the edge between $E_2$ and $E_3$. Since a directed path from $D_2$ through $E_3$ and $E_1$ to $D_0$ can not exist by Lemma 14, we deduce that the edge between $E_3$ and $E_1$ is non-positive for $E_3$. On the one hand we see from Lemma 8 that

$$\sum_{D_2}\tau + \sum_{D_3}\tau + \sum_{E_3}\tau + \sum_{E_4}\tau = \pi + \pi + \pi/3 + 2\pi/3 = 3\pi.$$

The inner edges of $D_2 \cup E_3 \cup D_3 \cup E_4$ occur once from each side in the sum and hence we need only sum the outer edges. But the positive edges with $E$ together contribute less than $\pi/3$, which leaves an impossibly large total curvature, $2\pi + 2\pi/3$, for the

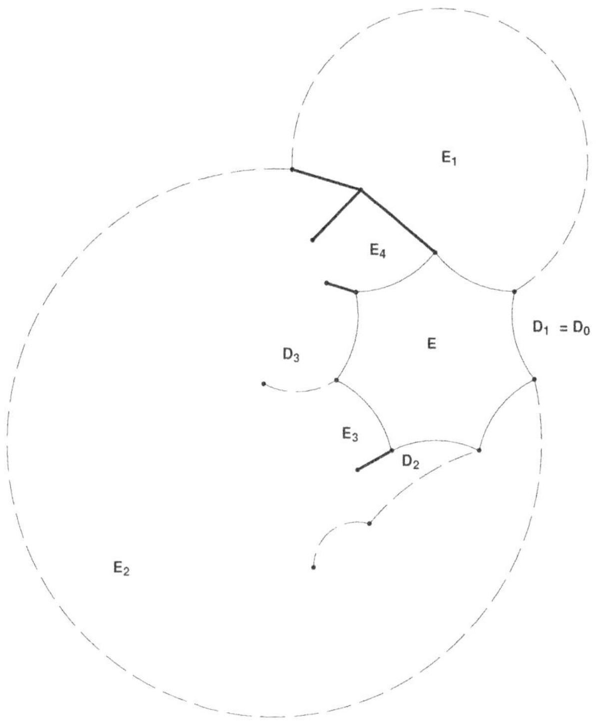


**Figure 7m3aiv.**

single edge between $E_1$ and $E_4$. At the junctions of $E_1$, $E_2$, and $E_3$ or the junction of $E_1$, $E_3$, and $E_4$ there could be another dead spaces, but then, adding the dead space(s) to the union, the argument would be similar.

If $E_{1,3}$ ends on $D_3$ we get the configuration of Figure 7m3iiB. As in the immediately preceding argument, the edge between $E_1$ and $E_3$ must be non-positive from the perspective of $E_3$. But now the five-edged component $E_3$ has its only positive edge with $E$ which contributes too little total curvature for the configuration to be optimal by Lemma 8. Should there be an additional dead space at the junction of $E_1$, $E_2$, and $E_3$, consideration of the union of $E_3$ with that dead space yields a contradiction.

If $e_{1,2}$ meets $D_3$, then $E_3$ has exactly the four edges shown in Figure 7m3aiii with edges having positive or negative curvature as indicated. Thus the sum of the total curvatures of the edges around $E_3$ is too small for optimality, and $e_{1,2}$ can't meet $D_3$.

Suppose that $e_{1,2}$ meets $E_4$, Figure 7m3aiv. We don't know if $e_{2,4}$ ends on $D_3$, $E_3$, or $D_2$.

If $e_{2,4}$ ends on $D_3$, we are lead to Figure 7m3aivA. Clearly the four-sided component $E_3$ has its only positive edge with $E$, but that edge contributes too little for the total curvature sum to be the $2\pi/3$ required by Lemma 8 for optimality. A possible additional dead space at the junction of $E_1$, $E_2$, and $E_4$ would not effect the argument.

If $e_{2,4}$ ends on $E_3$, we are lead to Figure 7m3aivB. If $E_3 \rightarrow E_4$, then the edges of $E_4$ must have negative curvature with $E_1$ and $E_2$ in order that there be no paths from $D_3$ to $D_0$. See Figure 7m3aivB1. Since all the edges of $E_4$ except that with $E$

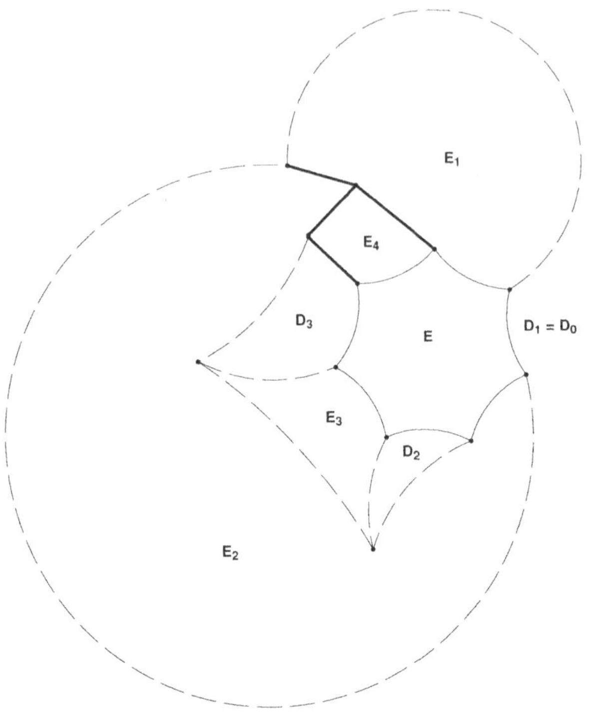


**Figure 7m3aivA.**

have non-positive curvature, the total curvature sum is less than the $\pi/3$ required by Lemma 8 for optimality. Should there be an additional dead space at the junction of $E_1$, $E_2$, and $E_4$, by adding that dead space to $E_4$ we would get a contradiction.

If the edge $e_{3,4}$ is not positive for $E_3$, then the contradiction comes from the total curvature sum of $E_3$. See Figure 7m3aivB2. The possible additional dead space where $E_1$, $E_2$, and $E_3$ meet is too far away to effect the argument. If there is an additional dead space at the juncture of $E_2$, $E_3$, and $E_4$, then the exterior curvature of $E_3 \cup D_2$ yields the contradiction.

If $e_{2,4}$ ends on $D_2$, we are lead to Figure 7m3aivC. We first suppose the edge between $D_2$ and $E_4$ is positive for $D_2$, Figure 7m3aivC1. The edge between $E_4$ and $D_3$ must have the same curvature. But to avoid a directed path from $D_2$ to $D_0$ the edges of $E_4$ with $E_1$ and $E_2$ must have non-positive curvature for $E_4$. Consider $E_1 \cup E_2$. By Lemma 8

$$\sum_{E_1} \tau + \sum_{E_2} \tau = 2\pi/3 + \pi/3.$$

But the sum is the same as the sum of the external edges of $E_1 \cup E_2$ all of which are non-negative and the two edges with $D_0$ contribute in excess of $3\pi$ to the sum, a contradiction. The possible additional dead space where $E_1$, $E_2$ and $E_4$ would strengthen the argument.

On the other hand if the edge between $D_2$ and $E_4$ is non-positive for $D_2$ we obtain Figure 7m3aivC2. We deduce that the edge of $E_3$ with $D_3$ is positive for $D_3$. $D_2$ must

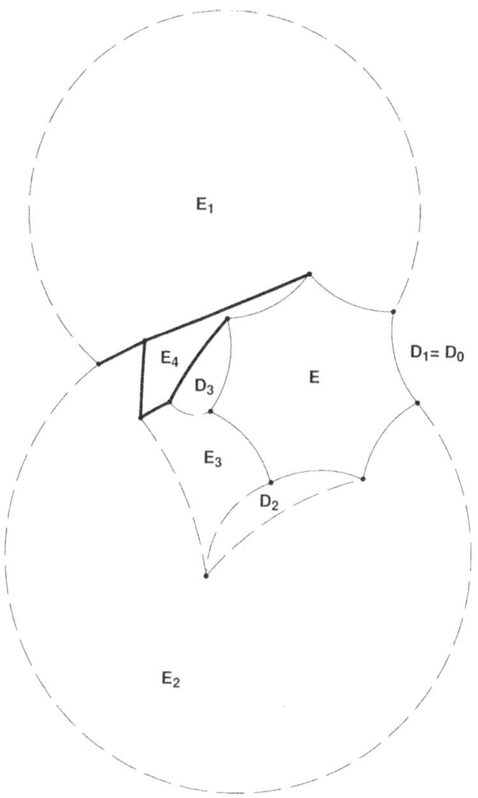


**Figure 7m3aivB.**

have a similar edge with $E_3$. Hence the edge between $E_3$ and $E_4$ is negative for $E_3$. But now we see that the total curvature sum of the edges $E_3$ is insufficient for optimality.

Thus $e_{1,2}$ does not meet $E_4$. But that exhausts all possible endpoints for $e_{1,2}$ so subcase 7m3a does not occur. We conclude that $D_1 = D_0$ is impossible.

**Subcase 7m3b:** $D_0 = D_2$

The following is a outline of the subcases of subcase 7m3b.

b) $D_0 = D_2$

- i) $e_{2,3} \Rightarrow E_4$
  - A) $e_{2,4} \Rightarrow E_1$
  - B) $e_{2,4} \Rightarrow D_1$
- ii) $e_{2,3} \Rightarrow D_3$
  - A) $e_{2,3*} \Rightarrow E_4$
    - 1) $e_{2,4} \Rightarrow E_1$
    - 2) $e_{2,4} \Rightarrow D_1$
  - B) $e_{2,3*} \Rightarrow E_1$

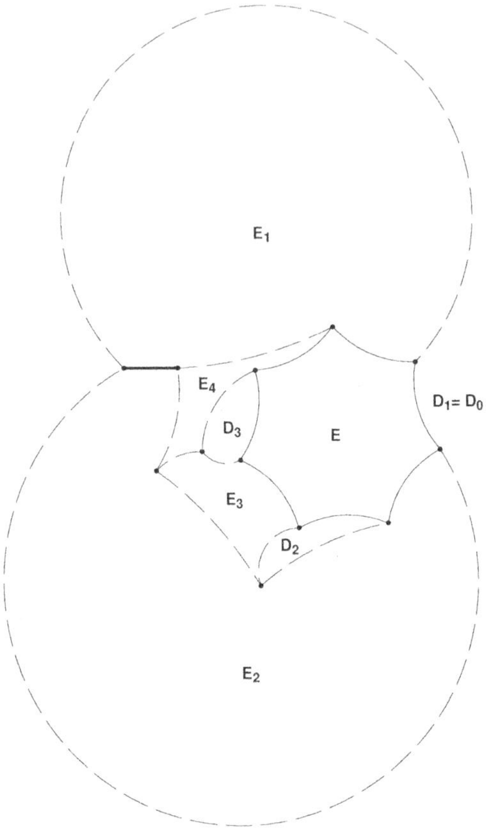


**Figure 7m3aivB1.**

We turn our attention to $D_2 = D_0$, Figure 7m3b. Again we shall need to consider the different options for the end of the edge $e_{2,3}$, separating $E_2$ and $E_3$. By symmetry we need only consider that it is on $E_4$ or $D_3$. We consider first when it ends on $E_4$, Figure 7m3bi. The edge $e_{2,4}$ might end on $E_1$ or $D_1$. Consider first that it ends on $E_1$, Figure 7m3biA. Clearly the edges of $D_3$ and $D_1$ must curve the way indicated. Then, in order to avoid a path joining either $D_1$ or $D_3$ to $D_0$ the edges of $E_4$ and $E_1$ with $E_2$ and $E_3$ must have non-positive curvature (shown negative). But then considering the sum of the total curvatures of $E_2 \cup E_3$ by the usual two methods yields a contradiction. There might possibly be other dead space where $E_2$, $E_3$, and $E_4$ meet, or where $E_1$, $E_2$, and $E_4$ meet, but that does not negate the contradiction. Now suppose $E_{2;4}$ ends on $D_1$, Figure 7m3biB. We first note the curvatures of $D_1$ and $D_3$ with $E_2$ and $E_3$, respectively. We then deduce the curvature of the edge between $D_3$ and $E_4$ which determines the curvature of the edge between $E_4$ and $D_1$. The vertices of$D_1$ and $D_3$ determine the curvatures of the edges of $E_4$ with $E_3$ and $E_2$. Now consider the sum of total curvatures on $E_2 \cup E_3$ all the outer edges of which have positive curvature. Those with $D_0$ have total curvatures summing to more than $3\pi$, too large for a two five-sided components. Again there could be another dead space where the three $E_i$ meet,

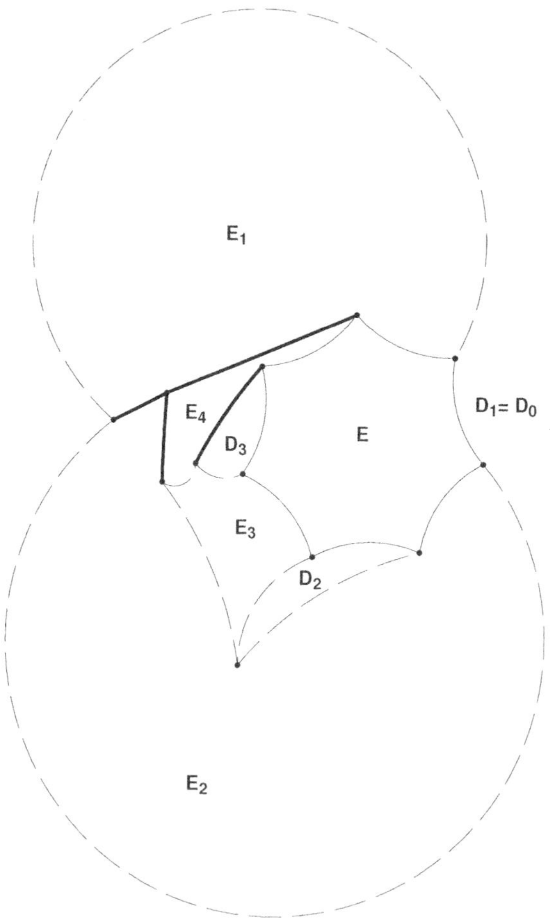


**Figure 7m3aivB2.**

but that won't materially effect the argument. We deduce that the edge $e_{2,3}$ can not meet $E_4$.

We now turn to the situation where $e_{2,3}$ meets $D_3$, Figure 7m3bii. We do not know whether the edge $e_{2,3*}$ ends on $E_4$ or $E_1$. It can not end on $D_1$, since then two dead spaces would meet. If $e_{2,3*}$ ends on $E_4$, Figure 73biiA, we still do not know whether the edge $e_{2,4}$ ends on $E_1$ or $D_1$. If it ends on $E_1$ we obtain Figure 7m3biiA1. Since $E_3$ and $E_2$ have positive edges with $D_0$, their edges with the other dead spaces are shown with correct curvature sign. The sum of the total curvatures for $D_3$ forces the curvature sign of the edge between $D_3$ and $E_4$, which forces the sign of the curvature of the edge between $E_2$ and $E_4$. But now $E_2 \cup E_3$ has all outer edges of positive curvature and the two with $D_0$ have curvature exceeding $3\pi$ which is a contradiction to the total curvature sum of $\pi/3$ required by Lemma 8 for optimality. The possible additional dead space where $E_1$, $E_2$, and $E_4$ would only strengthen the argument.

If $e_{2,4}$ meets $D_1$, we get the configuration of Figure 7m3biiA2, in which the total curvature sum of $E_2 \cup E_3$ ia again excessive.

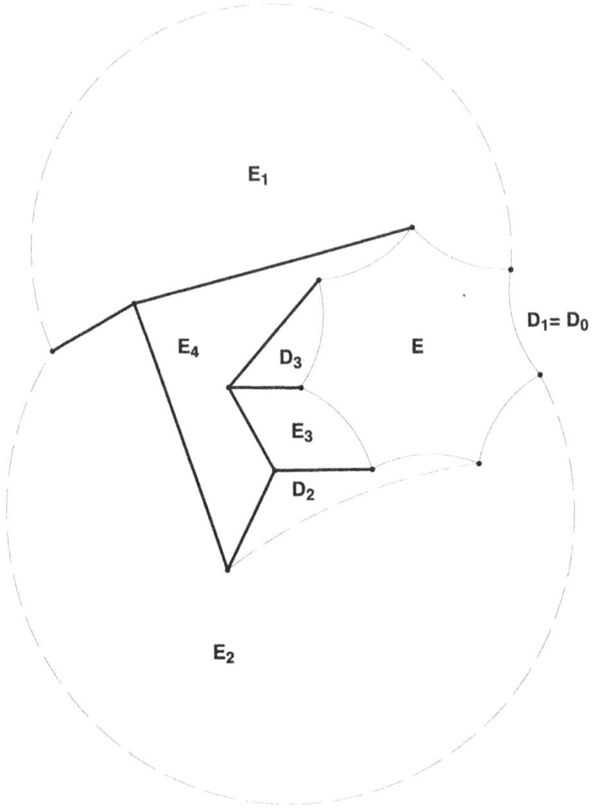


**Figure 7m3aivC.**

If $E_{2,3*}$ meets $E_1$ we obtain the configuration of Figure 7m3biiB. We note first that since $E_2$ has positive curvature with $D_0$ it must also with $D_1$ and $D_3$, which forces the edge of $D_1$ and $E_1$ to curve as shown, which in turn requires the edge between $E_1$ and $E_2$ to be positive for $E_2$. We now get a contradiction from considering the total curvature sum of the edges of $E_1 \cup E_2$.

Subcase 7m3b can not occur in an optimal network.

Case 7m3 can not occur in an optimal network.

**Case 7m4:**

In this case we have the same two subcases, depending on whether $D_0 = D_1$ (which is symmetric to $D_0 = D_3$) or $D_0 = D_2$.

Subcase 7m4a: $D_0 = D_1$

The following is a outline of the subcases of subcase 7m4a.

a) $D_0 = D_1$

  i) $E_4 \| D_0$

    A) $E_2 \to D_0$
    B) $\neg(E_2 \to D_0)$

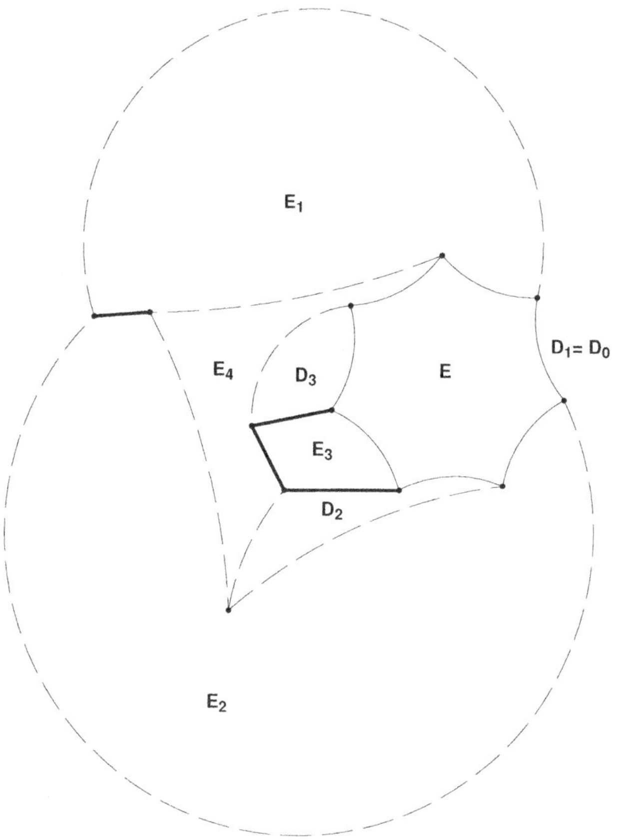


**Figure 7m3aivC1.**

ii) $E_3 \| D_0$

A) $e_{2,4} \Rightarrow D_2$

1) $E_4 \rightarrow D_0$
2) $\neg(E_4 \rightarrow D_0)$

B) $e_{2,4} \Rightarrow E_3$

1) $E_2 \rightarrow D_0$
  $\alpha$.) $E_3 \rightarrow D_4$ and $\neg(E_4 \rightarrow E_2)$
2) $\neg(E_2 \rightarrow D_0)$

C) $e_{2,4} \Rightarrow D_3$

1) $E_2 \rightarrow D_0 \quad E_4 \rightarrow D_0$
2) $E_1 \rightarrow D_0 \quad E_4 \rightarrow D_0$
3) $E_1 \rightarrow D_0 \quad E_2 \rightarrow D_0$

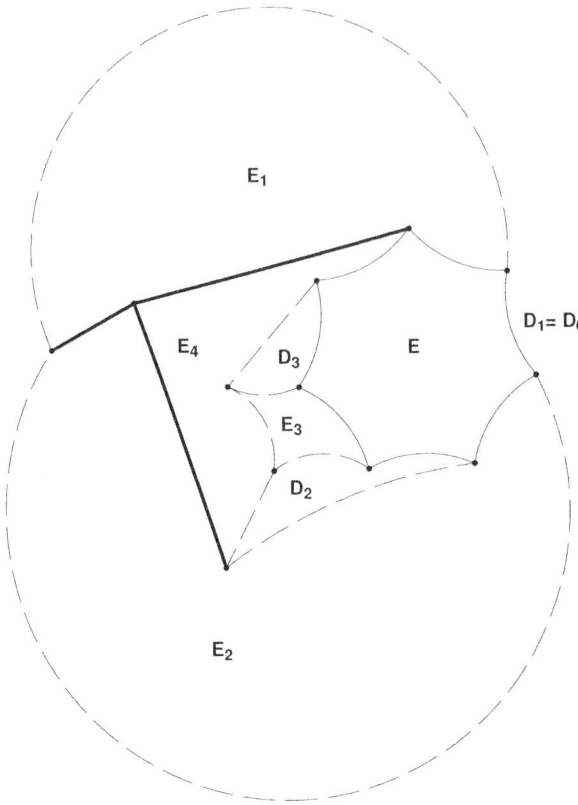


**Figure 7m3aivC2.**

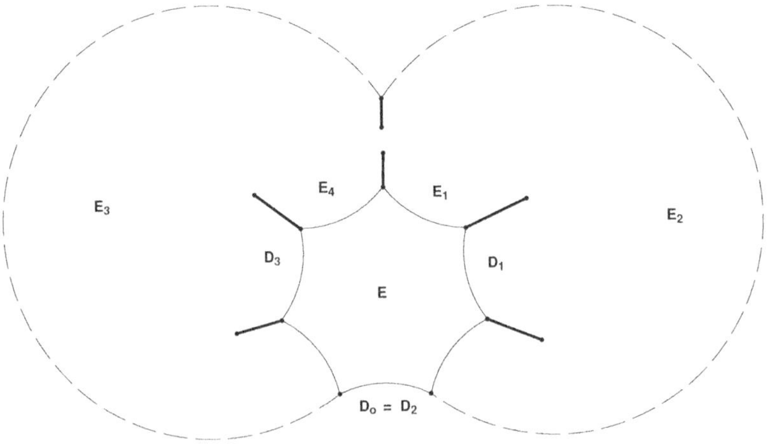


**Figure 7m3aivC2.**

In this case $E$, $E_1$, and $E_2$ meet $D_0$ and exactly one of $E_3$ and $E_4$ does. See Figure 7m4a. If it is $E_3$, Figure 7m4ai, then the only edges of $D_2$ are with $E$, $E_2$, and $E_3$. Hence not both $E_2$ and $E_3$ can have positive curvature with $D_0$. If $E_2$ has positive

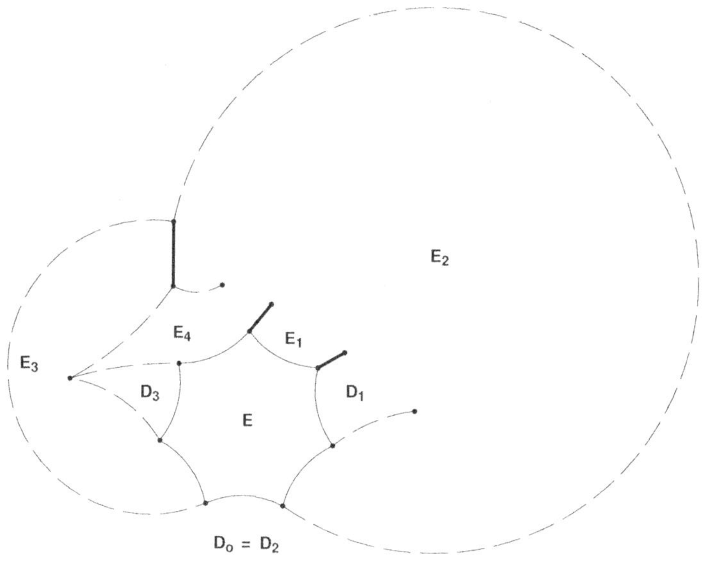


**Figure 7m3bi.**

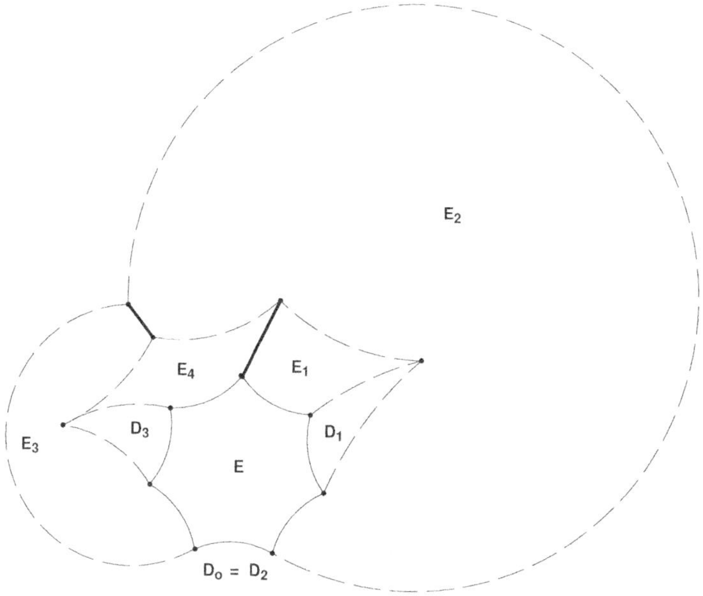


**Figure 7m3biA.**

curvature with $D_0$, Figure 7m4aiA, then the total curvature sum of the edges with $D_0$ is $3\pi$. This implies that the total curvature of the edge between $E_2$ and $D_0$ exceeds $\pi$, which leads to an impossibly large total curvature for the edges of $E_2$, all of which have positive curvature. If $E_3$ has positive curvature with $D_0$, we obtain Figure 7m4aiB. Both $E_3$ and $E_1$ have positive curvature with $D_0$; $E_2$ has negative curvature with $D_0$.

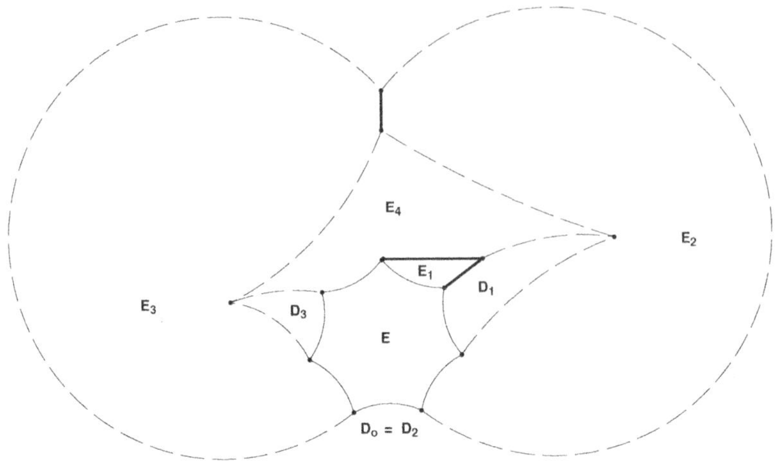


**Figure 7m3biB.**

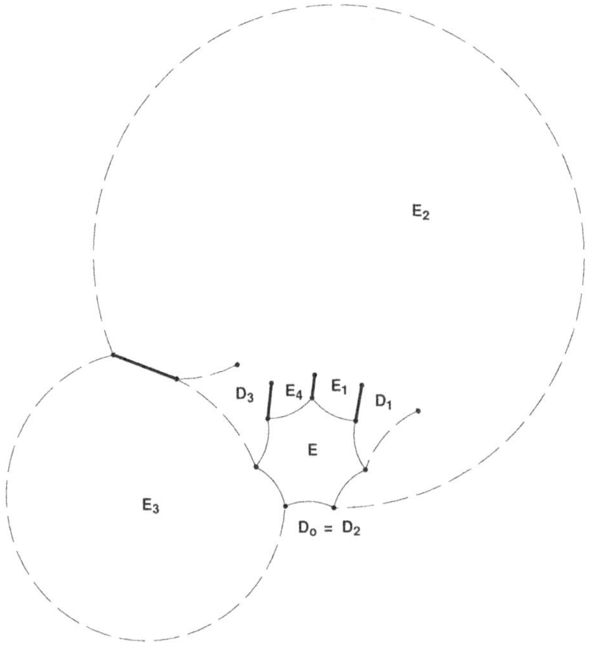


**Figure 7m3bii.**

We see that the only edge of $E_2$ with positive curvature is the one with $E$, so by Lemma 8, $E_2$ has insufficient total curvature sum for a four-sided component in an optimal configuration. Thus $E_3$ can not be the other component meeting $D_0$.

If $E_4$ meets $D_0$, we obtain Figure 7m4aii. The more detailed configuration depends on whether the edge $e_{2,4}$ between $E_2$ and $E_4$ has its other end on $D_2$, $E_3$, or $D_3$. If $e_{2,4}$ ends on $D_2$ we obtain Figure 7m4aiiA. If $E_4$ has positive curvature with $D_0$, Figure 7m4aiiA1, then it has positive curvature with all other dead spaces it meets.

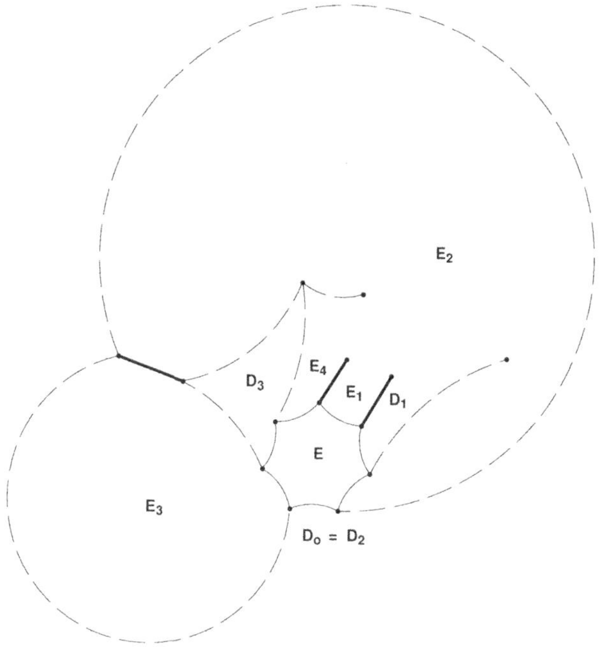


**Figure 7m3biiA.**

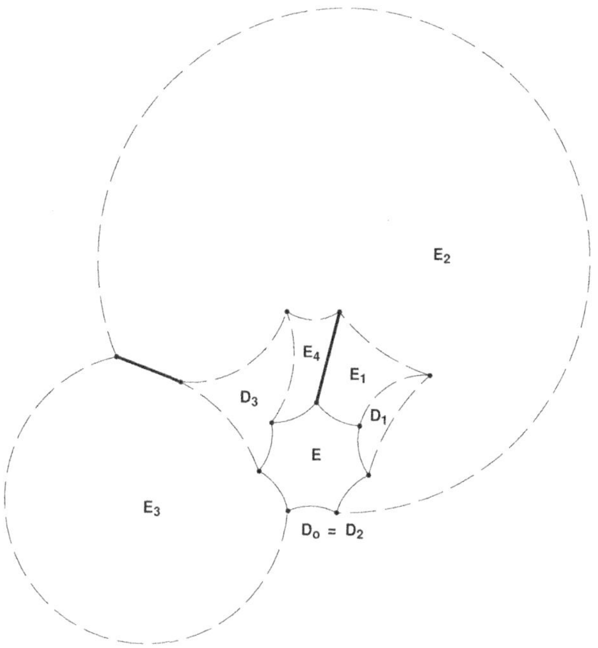


**Figure 7m3biiA1.**

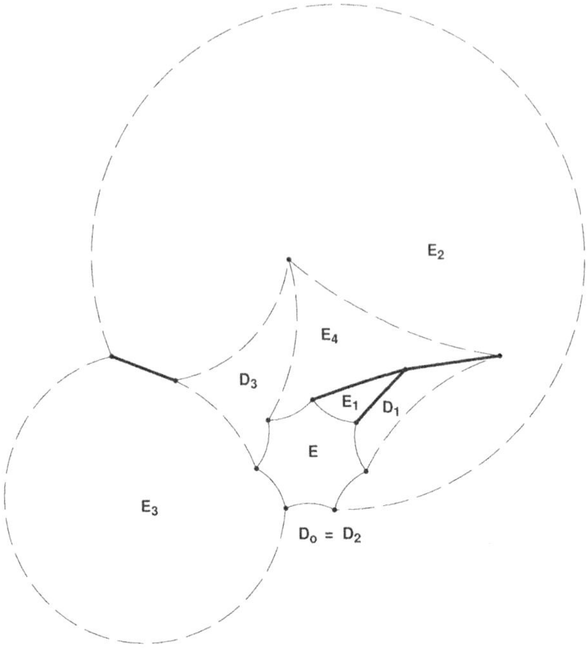


**Figure 7m3biiA2.**

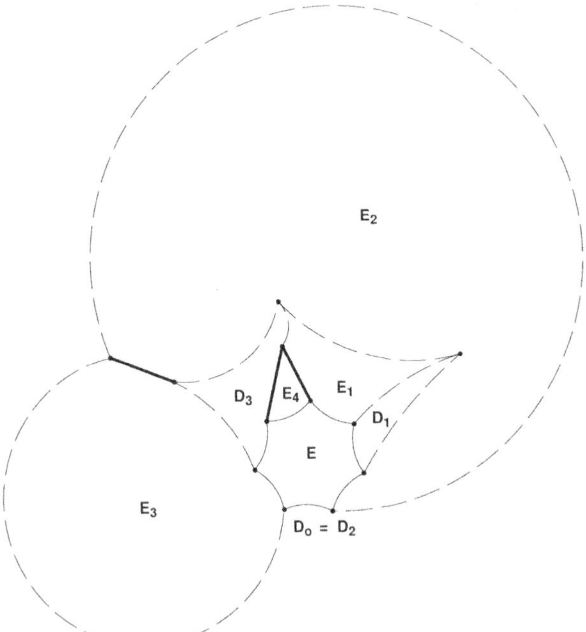


**Figure 7m3biiB.**

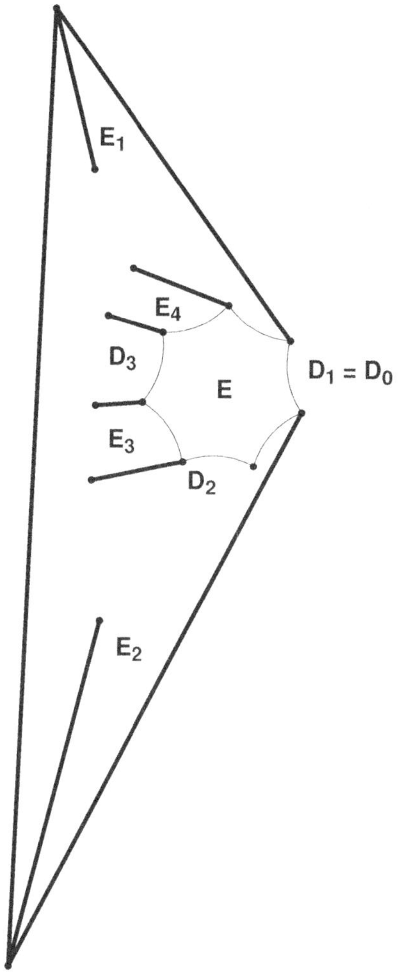


**Figure 7m4a.**

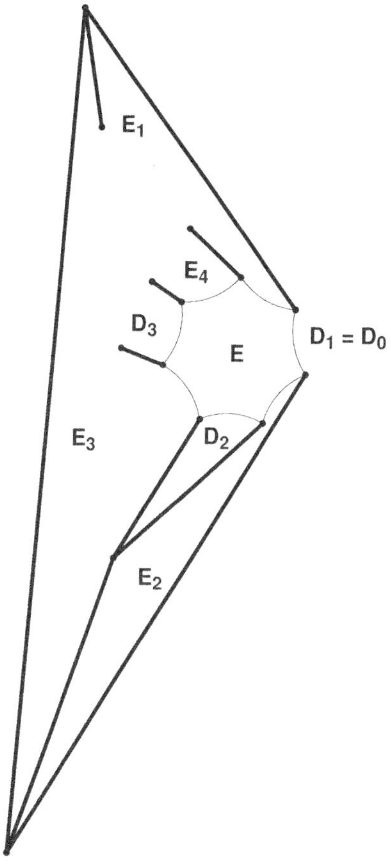


**Figure 7m4ai.**

Since $D_3$ is three-edged its third edge must have positive curvature into $E_3$. But then the edge separating $E_3$ from $E_4$ must also be negative for $E_3$. Since the edges of $E_3$ all have negative curvature except for the edge on $E$ which has small total curvature, the total curvature sum for $E_3$ is insufficient for optimality. Thus $E_4$ can not have positive curvature with $D_0$, but two of the regions meeting $D_0$ must have positive curvature, thus we are left to consider when $E_1$ and $E_2$ have positive curvature with $D_0$ and $E_4$ does not, Figure 7m4aiiA2. The curvature of $E_1$ and $E_4$ with $D_0$ determine that the curvature of the edge between them is positive for $E_1$. Thus all the sides of $E_1$ have positive curvature and the edge with $D_0$ has total curvature exceed $4\pi/3$, which make the total curvature sum of $E_1$ too large for optimality. We conclude that $e_{2,4}$ can not end on $D_2$.

We now consider when $e_{2,4}$ ends on $E_3$, Figure 7m4aiiB. If the edge of $E_2$ with $D_0$ has positive curvature for $E_2$, we quickly derive the curvatures shown in Figure 7m4aiiB1.

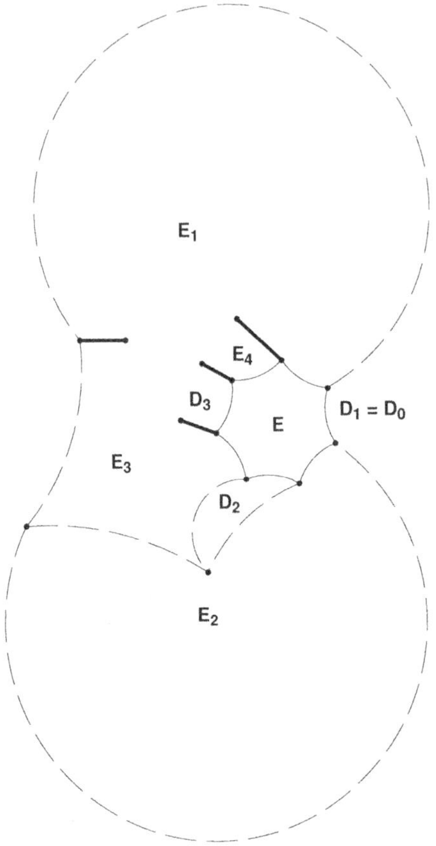


Figure 7m4aiA.

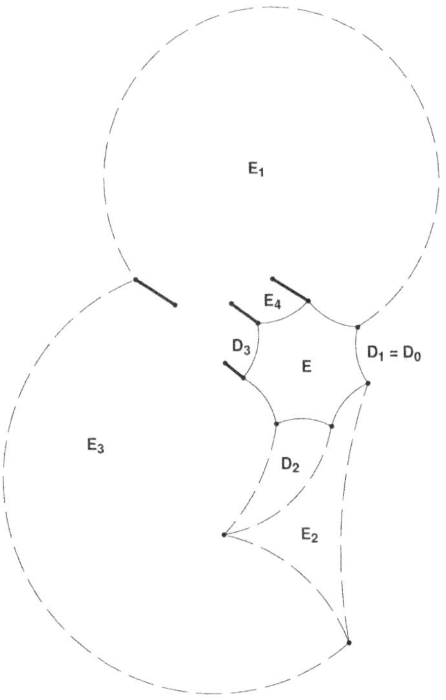


Figure 7m4aiB.

If the edge separating $E_4$ and $E_3$ does not have positive curvature for $E_3$ the total curvature sum of $E_3$ would be insufficient. If that edge does have positive curvature, the edge separating $E_2$ from $E_4$ would have positive curvature from $E_2$, Figure 7m4aiiB1a. A further consequence would be that the edge separating $E_4$ from $D_3$ would have positive curvature from the perspective of $D_3$. This would force a non-positive curvature from the perspective of $E_4$ onto the edge separating $E_4$ from $D_0$. Consequently the total curvature of the edge between $D_0$ and $E_2$ would exceed $4\pi/3$, so that the total curvature sum of $E_2$ would be excessive for optimality. A possible additional dead space at the junction of $E_2$, $E_3$, and $E_4$ does not negate the contradiction.

So the edge separating $E_2$ from $D_0$ has non-positive curvature from the perspective of $E_2$. See Figure 7m4aiiB2. It follows that both $E_1$ and $E_4$ have positive curvature with $D_0$ and the sum of the total curvatures of these two edges exceeds $10\pi/3$. From the curvatures of $E_2$ and $E_4$ with $D_0$ we deduce their curvatures with $D_2$ and $D_3$ respectively. But then the edge of $D_3$ with $E_3$ must be as shown, which shows that the edge of $E_4$ with $E_3$ is as shown. Finally, the vertex of the edge $e_{2,4}$ on $D_0$ determines that the curvature of $e_{2,4}$ is positive for $E_4$. But now we realize that all the exterior edges of $E_1 \cup E_4$ have positive curvature which means that the sum of the total curvatures

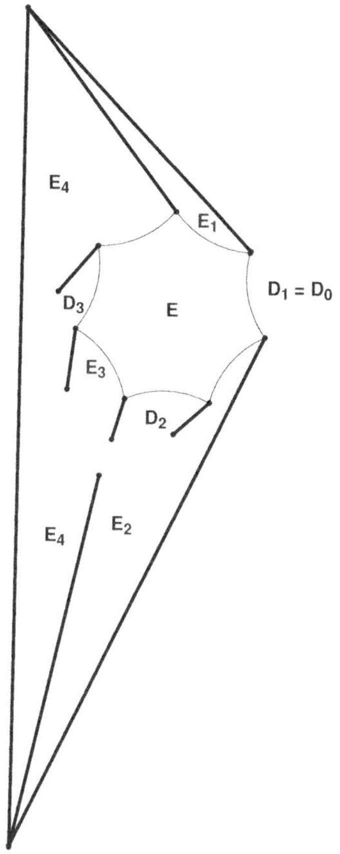


**Figure 7m4aii.**

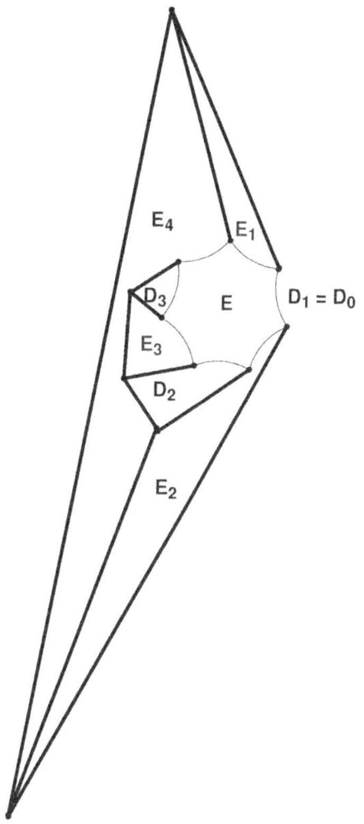


**Figure 7m4aiiA.**

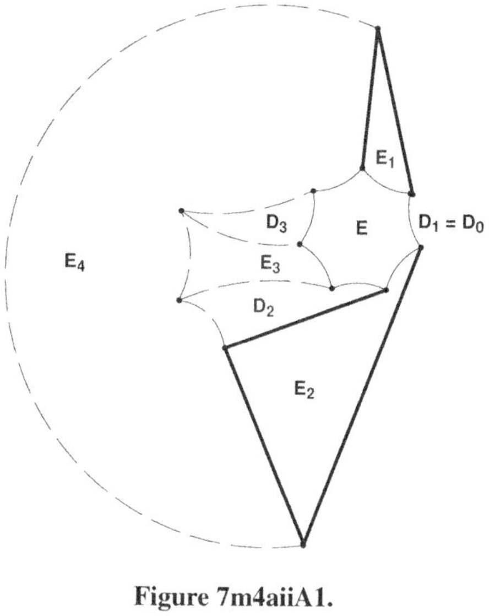


**Figure 7m4aiiA1.**

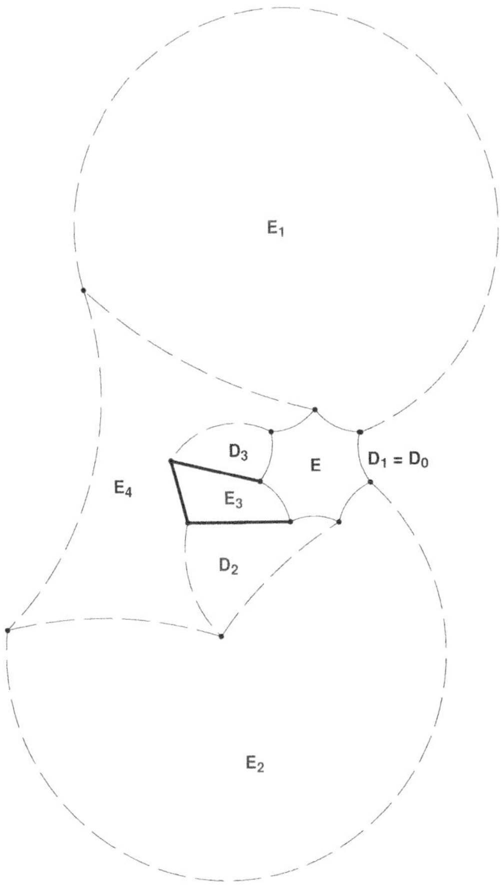


**Figure 7m4aiiA2.**

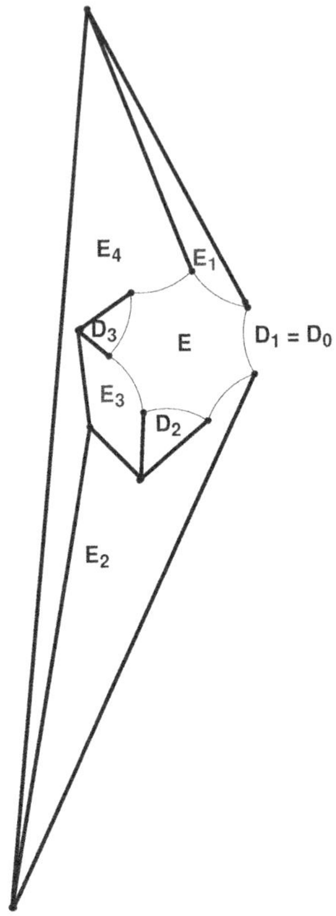


**Figure 7m4aiiB.**

exceeds $10\pi/3$, which contradicts optimality. Should there be a dead space at the conjunction of $E_2$, $E_3$, and $E_4$, the contradiction becomes more apparent.

Thus we are left to consider the possibility that $e_{2,4}$ ends on $D_3$. See Figure 7m4aiiC. Suppose both $E_2$ and $E_4$ can have non-negative curvature with $D_0$. See Figure 7m4aiiC1. We quickly deduce the curvature of the edges of $E_2$ and $E_4$ with the other dead spaces. This forces $E_3$ to have negative curvature edges with the dead spaces it meets. The vertices of those dead spaces force the curvature of the final edge of $E_3$. Since the only positive edge for $E_3$ is its edge with $E$, it does not satisfy the conditions for optimality.

Thus either precisely $E_1$ and $E_2$ or precisely $E_1$ and $E_4$ have positive curvature with $D_0$. We consider first $E_1$ and $E_4$ yielding the positive curvature with $D_0$, Figure 7m4aiiC2. By examining the vertex of $e_{2,4}$ with $D_0$ we obtain the curvature of that edge. We also know the curvature of the edge between the dead space $D_3$ and $E_4$ must be the same as with the edge on the free space $D_0$. This leads to all external edges of $E_1 \cup E_4$ having positive curvature and thus an impossibly large total curvature sum. We next consider when $E_1$ and $E_2$ are the pair with positive curvature with $D_0$. See

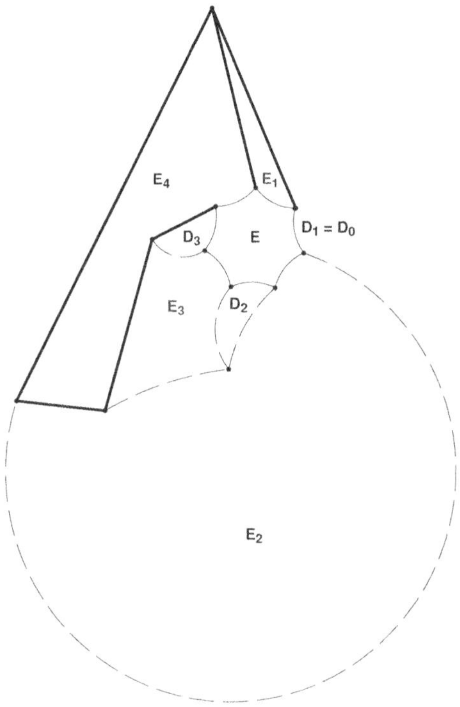


**Figure 7m4aiiB1.**

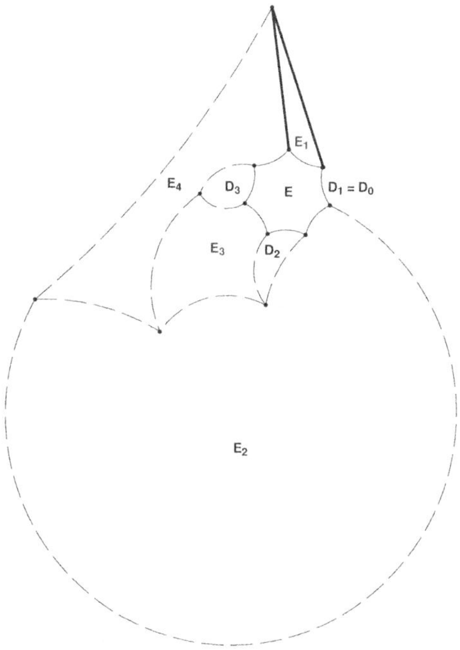


**Figure 7m4aiiB1a.**

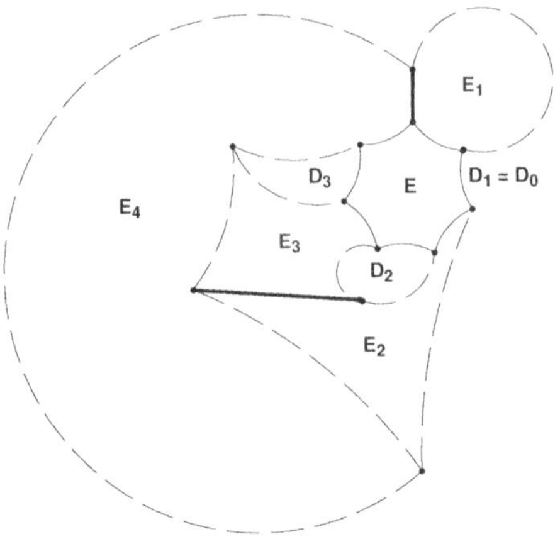


**Figure 7m4aiiB2.**

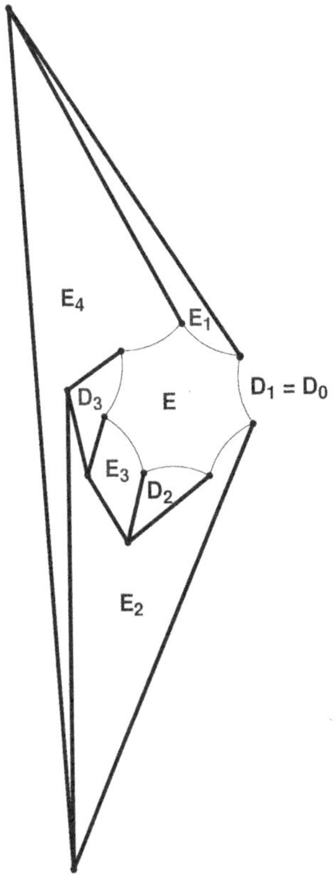


**Figure 7m4aiiC.**

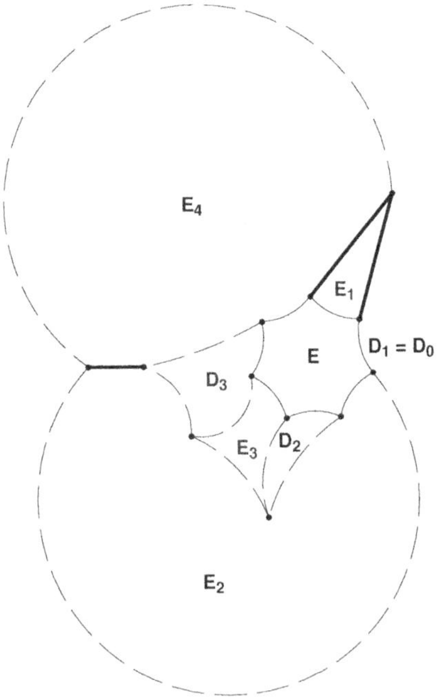


**Figure 7m4aiiC1.**

Figure 7m4aiiC3. By examining the vertex on $D_0$ of the edge separating $E_1$ and $E_4$ we determine its curvature as shown. But now the three-edged component $E_1$ has an impossibly large total curvature sum for optimality. Thus the variants of Figure 7m4aii are not possible.

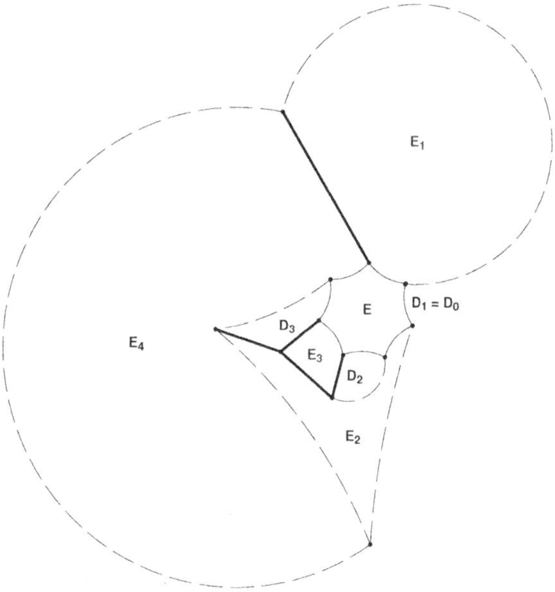


**Figure 7m4aiiC2.**

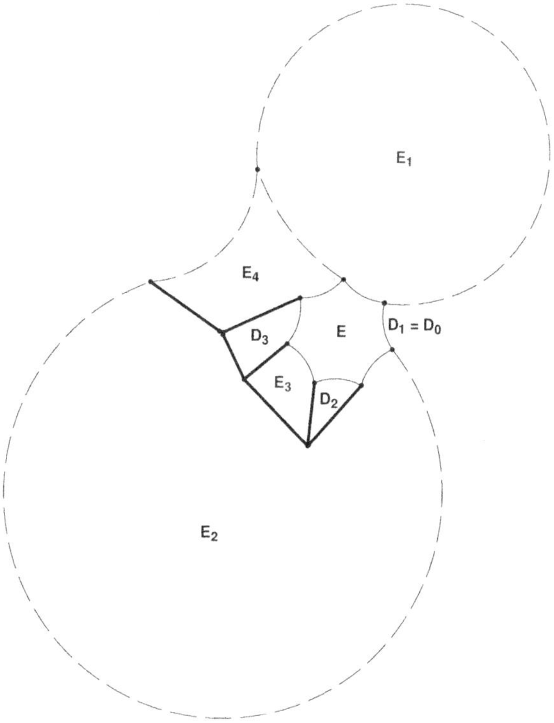


**Figure 7m4aiiC3.**

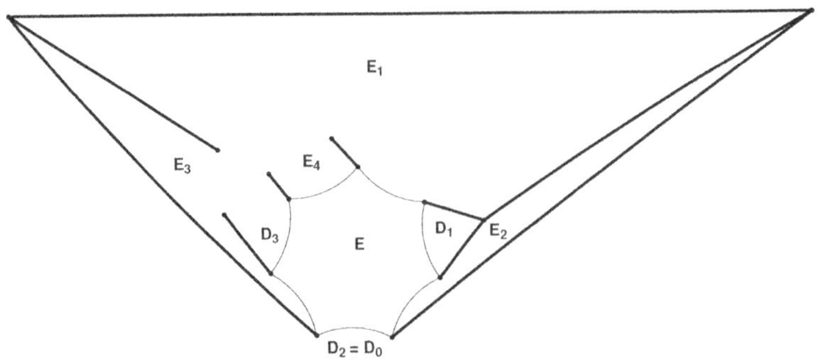


**Figure 7m4b.**

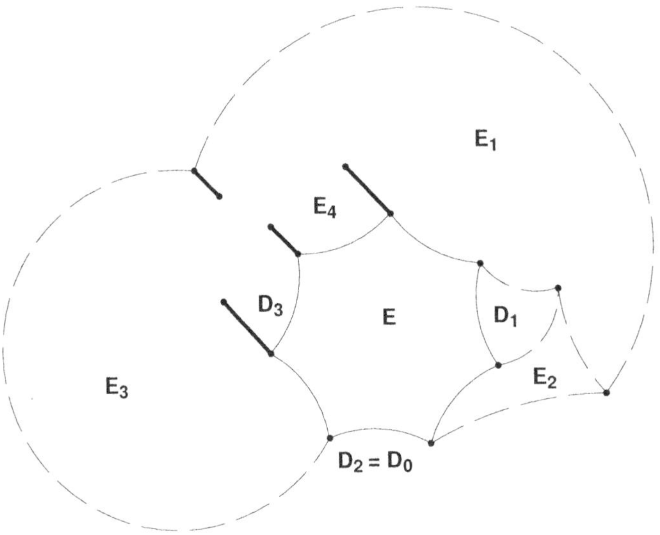


**Figure 7m4bi.**

Subcase 7m4a is impossible.

**Subcase 7m4b:** $D_0 = D_2$

The following is a outline of the subcases of Subcase 7m4b.

b) $D_0 = D_2$

i) $E_1 \to D_0 \quad E_3 \to D_0$
ii) $E_2 \to D_0 \quad E_3 \to D_0$

In this subcase one of $E_1$ or $E_4$ must meet $D_0$, but these are symmetric alternatives hence we need only consider that $E_1$ also meets $D_0$, Figure 7m4b. In this instance $E_1$ and $E_2$ can not both have non-negative curvature with $D_0$, since this would lead to the three-edged dead space $D_1$ having its only positive edge with $E$, but the total curvature of that edge is at most $\pi/3$. Thus the sum of the total curvatures of the edges of $D_1$

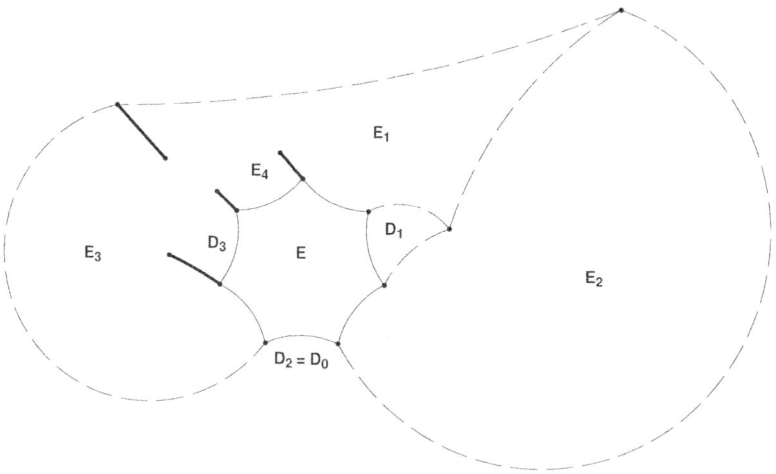


**Figure 7m4bii.**

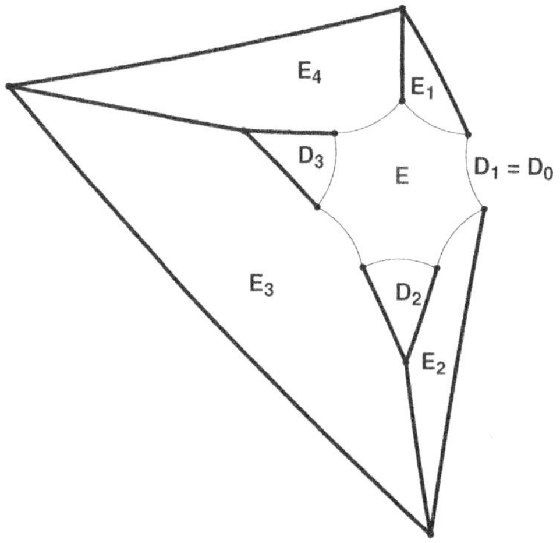


**Figure 7m5a.**

would be insufficient. Thus one of $E_1$ and $E_2$ meets $D_0$ with negative curvatures while the other and $E_3$ have positive curvature with $D_0$.

First, we consider when $E_1$ and $E_3$ have positive curvature with $D_0$ while $E_2$ has negative curvature, Figure 7m4bi. This determines the curvatures of the edges of $D_1$ But the edges of $D_1$ now determines the curvature of the edge between $E_1$ and $E_2$. We see that $E_2$ has four edges, the only edge with positive curvature is the edge with $E$ which has total curvature less than $\pi/3$. But a four-edged component must have sum of total curvatures equal to $2\pi/3$ to be optimal; hence $E_1$ may not have positive curvature with $D_0$.

Next, we consider when $E_2$ has $E_3$ positive curvature with $D_0$ while $E_1$ has negative curvature, Figure 7m4bii, then all the edges of $E_2$ have positive curvature and the edge

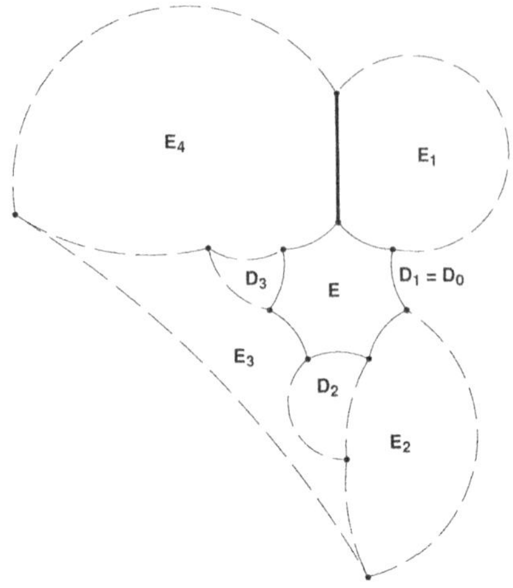


**Figure 7m5ai.**

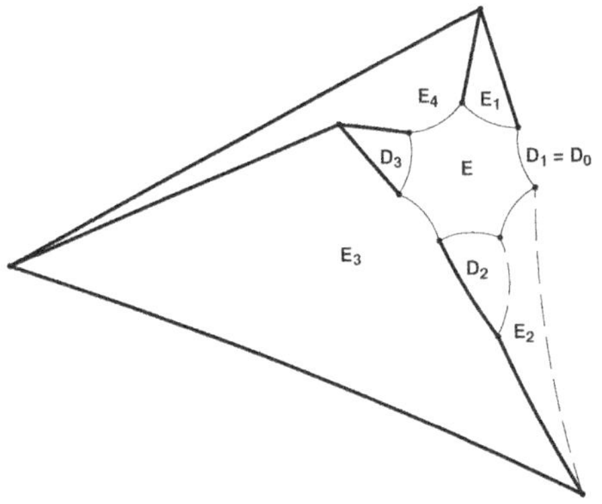


**Figure 7m5aii.**

with $D_0$ has total curvature of at least $4\pi/3$, which leads to too large a sum of total curvatures of its edges. This possibility can not occur. Consequently Subcase 7m4b is impossible.

It follows that Case 7m4 can not occur.

**Case 7m5:**

In this case each $E_i$, $i = 1, 2, 3, 4$ touches both $E$ and $D_0$. We consider two subcases according as $D_0$ is $D_1$ or $D_2$.

**Subcase 7m5a:** $D_0 = D_1$

The following is a outline of the subcases of Subcase 7m5a.

a) $D_0 = D_1$

  i) $\neg(D_0 \rightarrow E_2)$
  ii) $D_0 \rightarrow E_2$

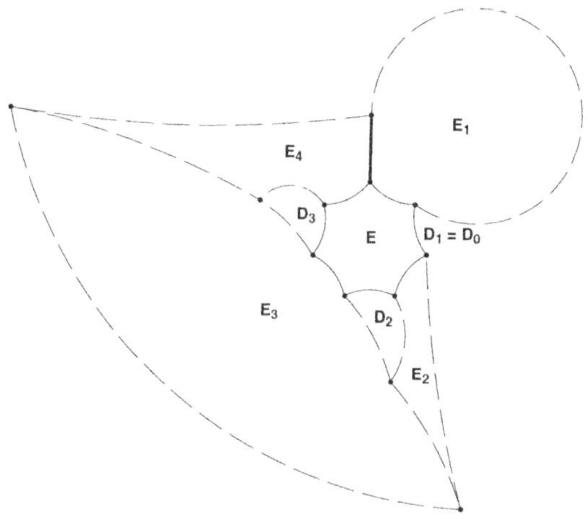


**Figure 7m5aiiA.**

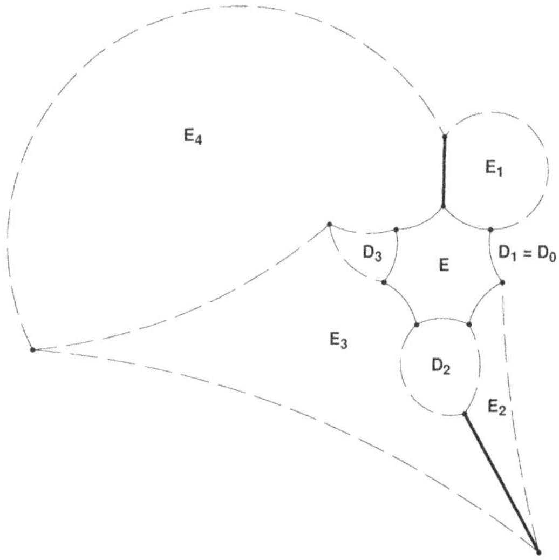


**Figure 7m5aiiB.**

A) $\neg(D_0 \to E_3)$
B) $D_0 \to E_3$

See Figure 7m5a.

We first suppose $E_2$ has a non-negative edge with $D_0$, Figure 7m5ai. We see that the edges of $D_2$ must be as shown, which forces the edge separating $E_2$ from $E_3$ to be positive from the perspective of $E_2$. Since $E_2$ has four-edges and no edges with negative curvature its edge with $D_0$ has total curvature less than $2\pi/3$ required by Lemma 8. The sum of the total curvatures of the edges of $D_0$ is $2\pi + 5\pi/3$, we deduce that the total curvatures of the edges of $E_1$ and $E_4$ with $D_0$ each exceed $\pi$ and sum to

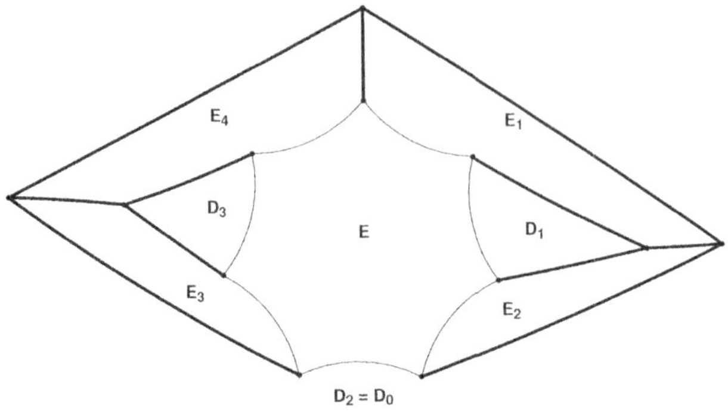


Figure 7m5b.

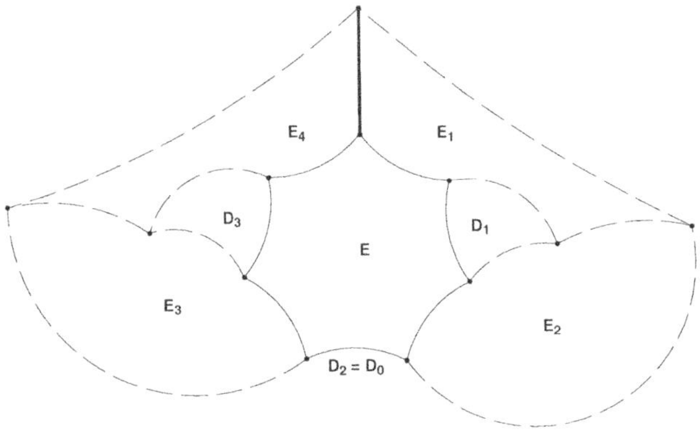


Figure 7m5bi.

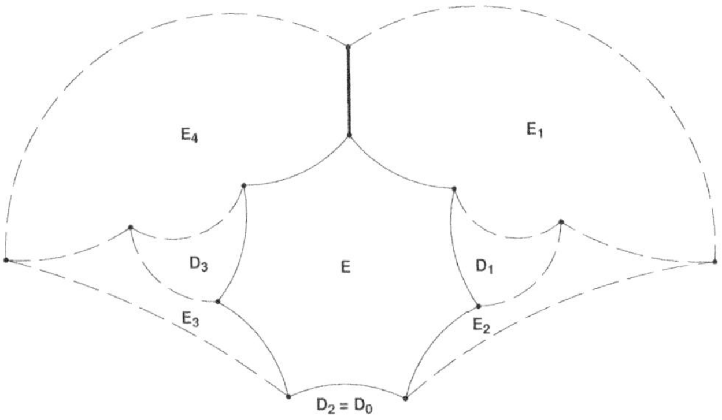


Figure 7m5bii.

more than $3\pi$. This enables us to determine the sign of the curvatures of all the edges, except the one indicated. To be optimal

$$\sum_{E_1} \tau + \sum_{E_4} \tau = \pi + \pi/3,$$

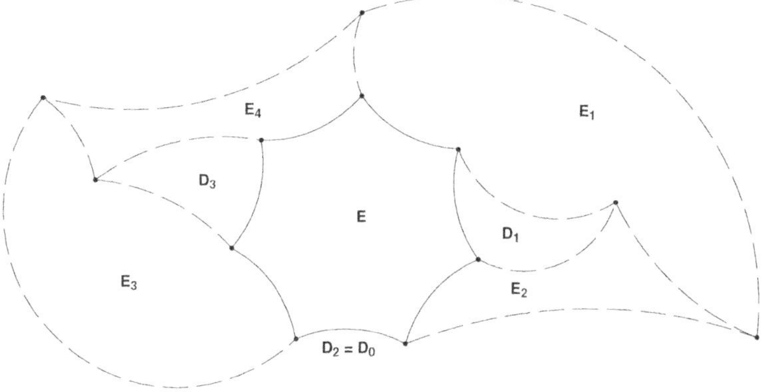


**Figure 7m5biii.**

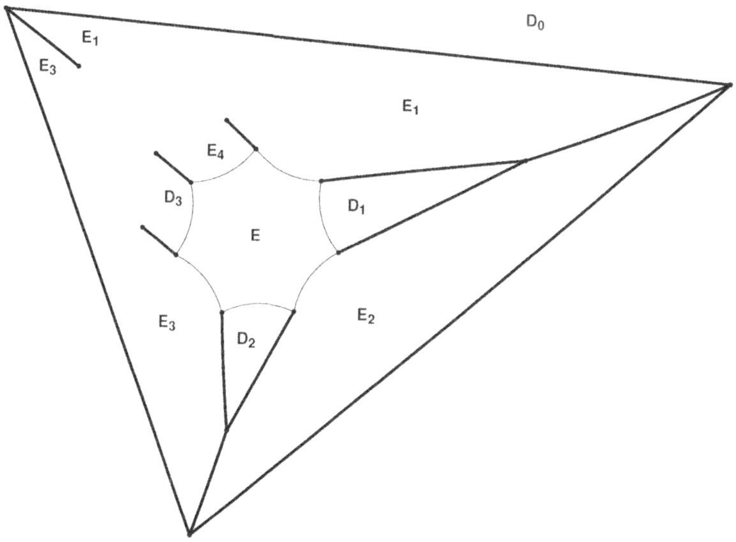


**Figure 7s3a.**

but the edges with $D_0$ contribute more than that. Consequently the edge of $E_2$ with $D_0$ has negative curvature from the perspective of $E_2$, Figure 7m5aii. We do not know the curvature of the edge between $D_0$ and $E_3$.

If the edge of $E_3$ with $D_0$ is not negative for $E_3$, Figure 7m5aiiA, then the edge separating $E_2$ from $E_3$ must be positive from the perspective of $E_3$. Thus each edges of $E_2$ has negative total curvature except the edge with $E$ which has total curvature less than $\pi/3$. So the sum of the total curvatures of the edges of $E_2$ is insufficient for optimality by Lemma 8.

We now suppose the edge of $E_3$ with $D_0$ is negative, Figure 7m5aiiB, in which case the edges of $E_1$ and $E_4$ with $D_0$ have positive curvature. By familiar arguments we deduce the indicated curvatures of Figure 7m5aiiB. The sum of the total curvatures of

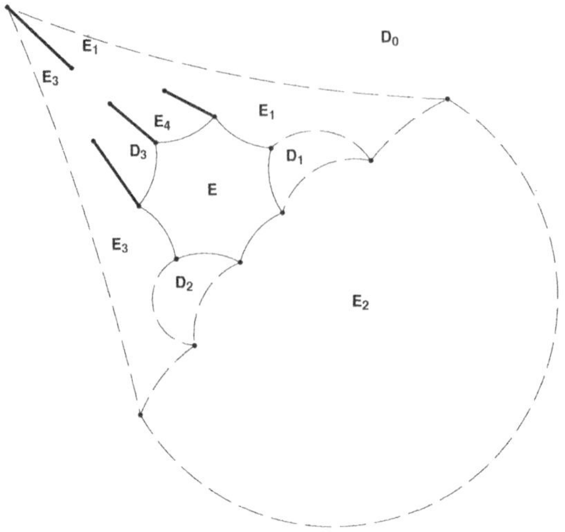


**Figure 7s3ai.**

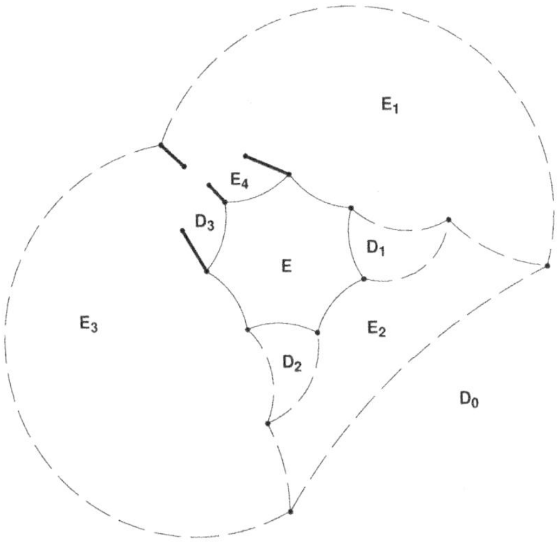


**Figure 7s3aii.**

the edges of $E_1$ and $E_4$ with $D_0$ must exceed $2\pi + 5\pi/3$ by Lemma 8. But this forces the total curvature sum of $E_1 \cup E_4$ to exceed the requirement for optimality.

Subcase 7m5a can not occur.

Subcase 7m5b: $D_0 = D_2$

The following is a outline of the subcases of Subcase 7m5b.

b) $D_0 = D_2$

i) $E_2 \to D_0 \quad E_3 \to D_0$
ii) $E_1 \to D_0 \quad E_4 \to D_0$
iii) $E_1 \to D_0 \quad E_3 \to D_0$

See Figure 7m5b.

At least two of the $E_i$ must have positive curvature with $D_0$, but not both of the pair $E_1$ and $E_2$ and not both of the pair $E_2$ and $E_4$. See Figure 7m5b. We consider the three possibilities: i) Both $E_2$ and $E_3$ have positive curvature. ii) Both $E_1$ and $E_4$ have positive curvature. iii) Both $E_1$ and $E_3$ have positive curvature. Possibility iii) is symmetric with the $E_4$ and $E_2$, so these are all the possibilities we need consider.

We first examine when both $E_2$ and $E_3$ have positive curvature with $D_0$, Figure 7m5bi. Since $E_2$ has positive curvature with $D_0$ it does with $D_1$ also. This forces the edge of $D_1$ to curve into $E_1$. The vertex of $D_1$ then determines the sign of the curvature of the edge between $E_2$ and $E_1$. But we now see the total curvatures of the edges of $E_2$ have too large a sum to be optimal.

We next examine when both $E_1$ and $E_4$ have positive curvature with $D_0$, Figure 7m5bii. Again starting with $E_1$ we deduce that the right-hand side of the figure is correct and the left- hand side is symmetric with it. The contradiction now comes from the total curvature sum of $E_2$ which has all non-positive edges, except the one with $E$ which contributes too little for an optimal configuration.

Now we examine when both $E_1$ and $E_3$ have positive curvatures with $D_0$, Figure 7m5biii. By, hopefully, now familiar techniques we deduce the correctness of the figure. The contradiction now comes from the sum of the total curvatures of $E_2$.

Subcase 7m5b dos not occur.

Case 7m5 dos not occur.

**Case 7s3:**

In this case $D_0$ meets all but one of the four components $E_i, i = 1, 2, 3, 4$. By the symmetries there are two subcases, the one $D_0$ does not meet might be one of the components $E_1$ or $E_4$, which are symmetric alternatives or one of $E_2$ or $E_3$ which are also symmetric alternatives. This leads to two subcases.

Subcase 7s3a: $E_4$ does not meet $D_0$

The following is a outline of the subcases of Subcase 7s3a.

a) $E_4 \| D_0$

   i) $\neg(D_0 \rightarrow E_2)$
   ii) $D_0 \rightarrow E_2$

See Figure 7s3a.

Since $E_i, i = 1, 2, 3$, all touch both $E$ and $D_0$, we deduce that $D_1$ and $D_2$ are both three-edged. If $E_2$ had non-negative curvature with $D_0$, Figure 7s3ai, then neither $E_1$ or $E_3$ could have positive curvature with $D_0$, since each has a three-edged dead space between it and $E_2$. However at least two of the edges of $D_0$ have positive curvature from the perspective of the components bounding it, Figure 7s3ai is impossible in an optimal network.

Thus $E_2$ has negative curvature with $D_0$ and $E_1$ and $E_3$ have large positive curvature with $D_0$. See Figure 7s3aii. Because of the curvatures of $E_i, i = 1, 2, 3$, we deduce the curvatures of the edges of $D_1$ and $D_2$ as shown. This forces the curvatures of the

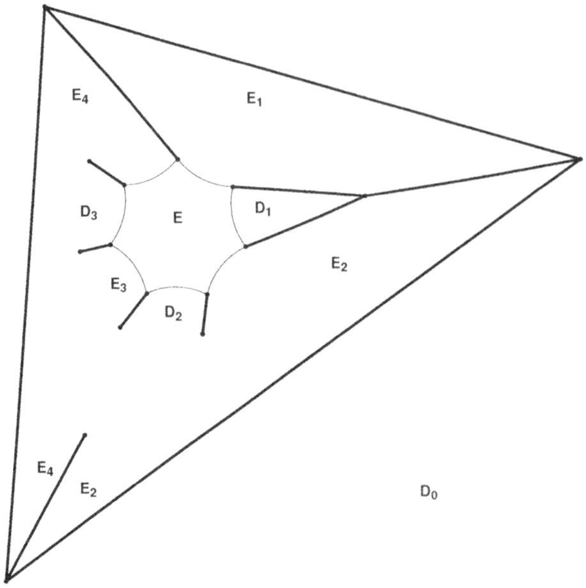


**Figure 7s3b.**

edges of $E_2$ with $E_1$ and $E_3$ to also be negative from the perspective of $E_2$. Consider the equation

$$\sum_{D_1} \tau + \sum_{E_2} \tau + \sum_{D_2} \tau = \pi + 0 + \pi,$$

where the right hand side comes from Lemma 8. But in summing those total curvatures, those internal to $D_1 \cup E_2 \cup D_2$ are summed from each side and cancel, while all the other edges except those with $E$ have negative curvature. But the edges of $E$ contribute less than $\pi/3$ so the sum is short of the $2\pi$ required for optimality. This subcase is impossible.

Subcase 7s3b: $E_3 \| D_0$

The following is a outline of the subcases of Subcase 7s3b.

b) $E_3 \| D_0$

  i) $\neg(E_2 \rightarrow D_0)$

    A) $e_{2,4} \Rightarrow D_2$
    B) $e_{2,4} \Rightarrow E_3$
    C) $e_{2,4} \Rightarrow D_3$

  ii) $E_2 \rightarrow D_0$

We know at least two of the three edges meeting $D_0$ have positive curvature with it, Figure 7s3b. Suppose $E_2$ had non-positive curvature with $D_0$, then both $E_1$ and $E_4$ would have positive curvature with $D_0$. This leads to the edge configuration of

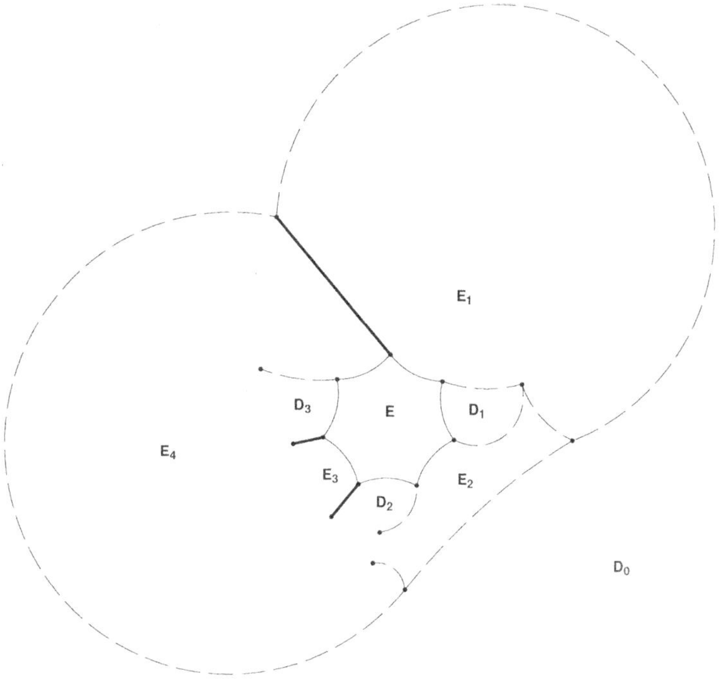


**Figure 7s3bi.**

Figure 7s3bi where the sign of the curvatures of the edges separating $E_2$ from $E_1$ and $E_4$ are determined by examining the endpoint on $D_0$. Further analysis depends on the location of the other endpoint of the edge $e_{2,4}$ which separates $E_2$ from $E_4$. The endpoint must be on one of $D_2$, $E_3$, or $D_3$.

We first suppose $e_{2,4}$ ends on $D_2$, Figure 7s3biA. By looking at the sum of the total curvatures of the edges $D_1 \cup E_2$ we see that the only external edges with positive curvature are on $E$, and those edges together contributes less than $\pi/3$ to the total curvature sum which therefore falls short of the required sum of $\pi$ according to Lemma 8.

We next consider when $e_{2,4}$ ends on $E_3$, Figure 7s3biB. Here the contradiction comes from considering the sum of the total curvatures of the edges of $E_1 \cup E_4$. All the external edges have positive curvatures and the total curvatures of the two edges with $D_0$ contribute more than $3\pi$. Thus the sum exceeds the amount required by Lemma 8 for optimality. Should there be an additional dead space where $E_2$, $E_3$, and $E_4$ meet, the essential argument is the same.

We now consider when $e_{2,4}$ ends on $D_3$, Figure 7s3biC. Once again $E_1 \cup E_4$ yields the contradiction to optimality as in the preceding paragraph.

We conclude $E_2$ must have positive curvature with $D_0$, but then via, $D_1$, we conclude $E_1$ has negative curvature with $D_0$, so $E_4$ must have positive curvature with $D_0$, which, in turn, forces the edge between $E_1$ and $E_4$ to have negative curvature for $E_1$, Figure 7s3bii. The sum of the total curvatures of the edges of $E_1$ must be $\pi/3$ to be optimal, but the only edge with positive curvature has total curvature less than $\pi/3$, a contradiction.

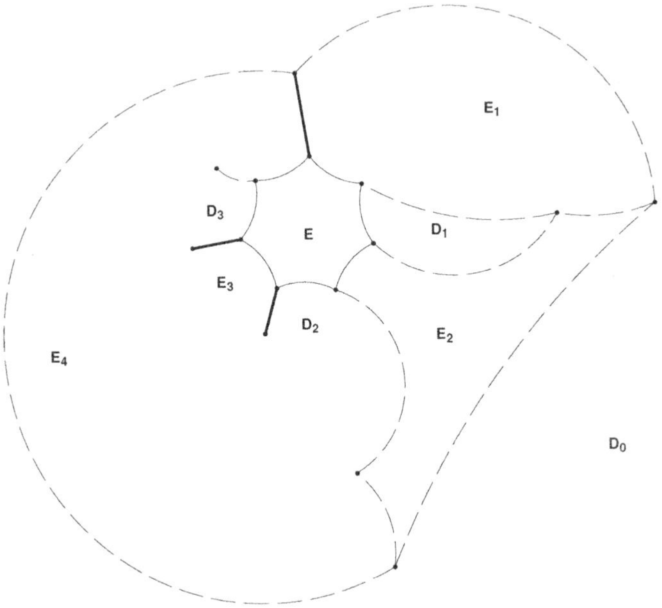


**Figure 7s3biA.**

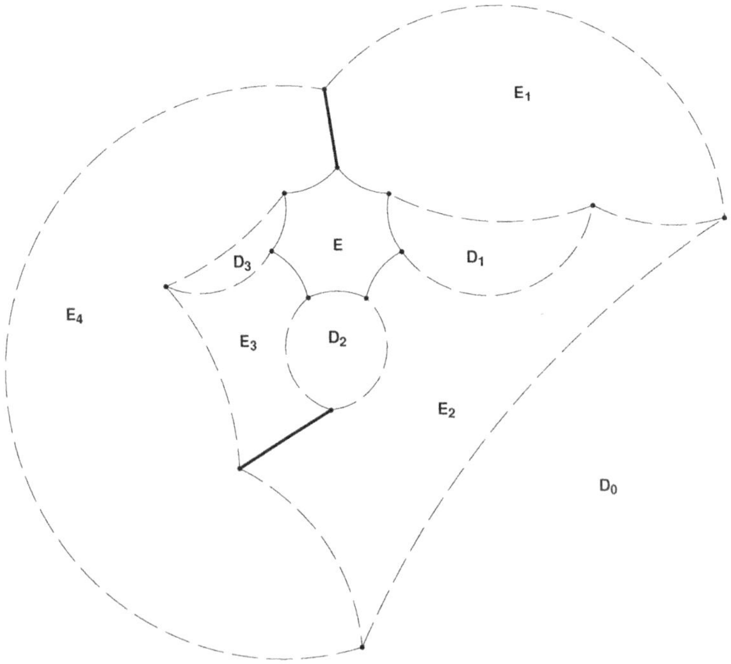


**Figure 7s3biB.**

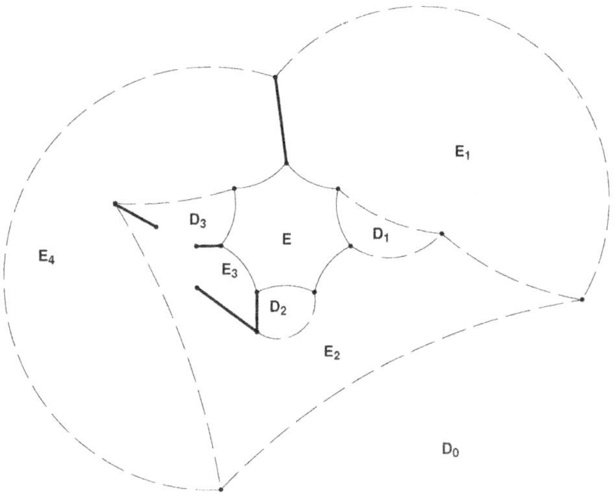


**Figure 7s3biC.**

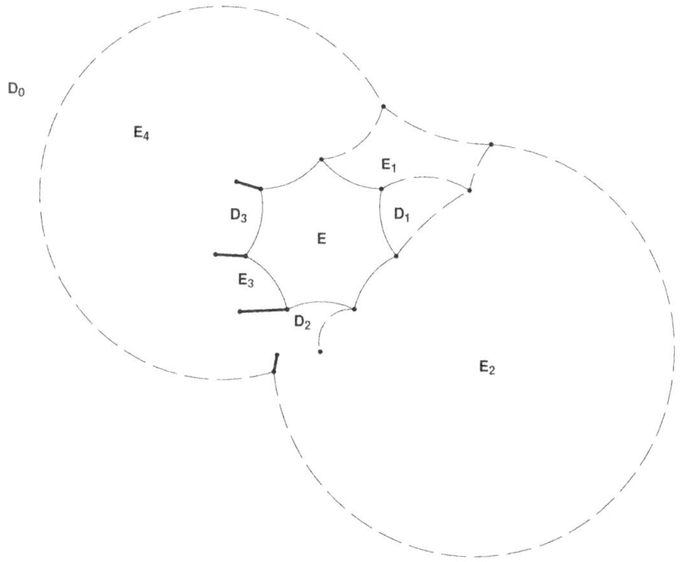


**Figure 7s3bii.**

Subcase 7s3b can not occur.
Case 7s3 can not occur.

**Case 7s4:**

In this case all the $E_i$ meet $D_0$, Figure 7s4.

If $E_2$ has positive curvature with $D_0$ then neither $E_1$ nor $E_3$ can. Thus $E_2$ and $E_4$ are the only possibility with $E_2$ having a positive curvature edge with $D_0$. The case where $E_3$ has a positive curvature edge with $D_0$ is symmetric. The other possibility is that $E_1$

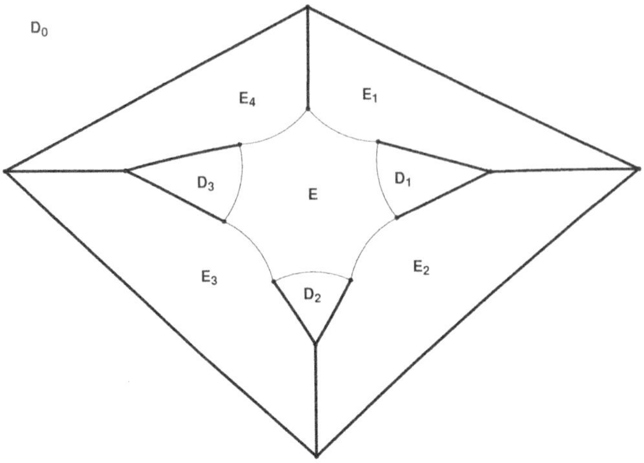


**Figure 7s4.**

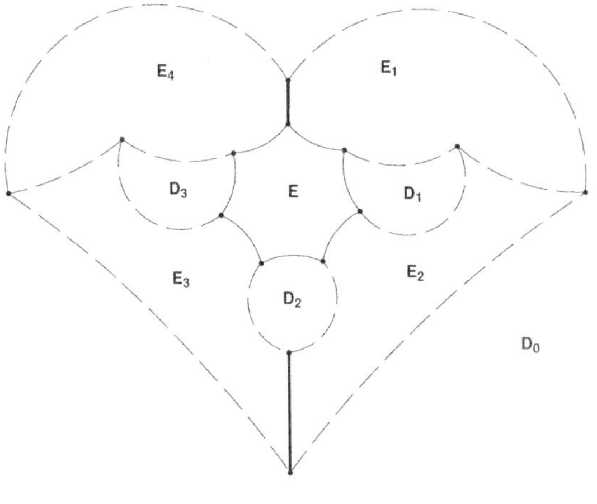


**Figure 7s4a.**

and $E_4$ are the two components with positive curvatures with $D_0$. Thus we consider two subcases.

Subcase 7s4a: Precisely $E_1$ and $E_4$ have positive curvatures with $D_0$

We may quickly deduce the correctness of the curvatures shown in Figure 7s4a. Once again, the total curvature of $E_1 \cup E_4$ will show this is not optimal. This subcase is impossible.

Subcase 7s4b: Precisely $E_1$ and $E_3$ have positive curvatures with $D_0$

In this case $E_3$ will have no edges of negative curvature. The curvatures of its edges with $E_2$ and $E_4$ are determined by their endpoints on $D_0$. Since $E_3$ has six edges its total curvature sum must be zero to be optimal. But the edge with $D_0$ has large positive total curvature which makes the sum too large. This subcase is impossible.

Case 7s4 can not occur.

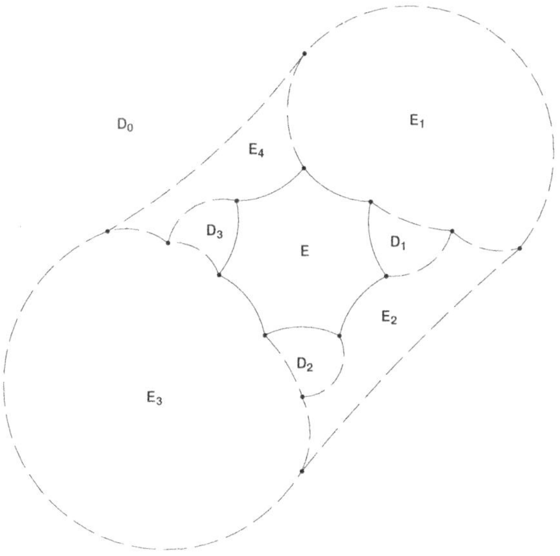


**Figure 7s4b.**

**Case 8m3:**

The following is a outline of the subcases of Subcase 8m3.

a) $e_{1,4} \Rightarrow D_2$
b) $e_{1,4} \Rightarrow E_3$

  i) $e_{1,3} \Rightarrow D_2$
  ii) $e_{1,3} \Rightarrow E_2$
  iii) $e_{1,3} \Rightarrow D_1$

c) $e_{1,4} \Rightarrow D_3$

  i) $e_{1,3*} \Rightarrow E_2$
  ii) $e_{1,3*} \Rightarrow E_3$

    A) $e_{1,3} \Rightarrow D_2$
    B) $e_{1,3} \Rightarrow E_2$
    C) $e_{1,3} \Rightarrow D_1$

Without loss of generality, we may assume $D_0 = D_4$. Since $D_0$ has three edges and clearly meets the three bounded components $E$, $E_1$, and $E_4$, we obtain Figure 8m3. We consider the possible other endpoint of the edge $e_{1,4}$ which separates $E_1$ from $E_4$. By symmetry we need only consider it ends on $D_2$, $E_3$ or $D_3$.

Suppose, by way of contradiction, that it ends on $D_2$, Figure 8m3a. Since the sum of the total curvatures of the edges of $D_0$ is $3\pi$ and the total curvature of its edge with $E$ is negative, each of the other two edges of $D_0$ must have total curvature greater than $\pi$ when viewed from the bounded component. This forces the directed path between $E_1$ and $D_1$ to go from $E_1$ into $D_2$. Since the sum of the total curvatures of $D_1$ is $\pi$, the

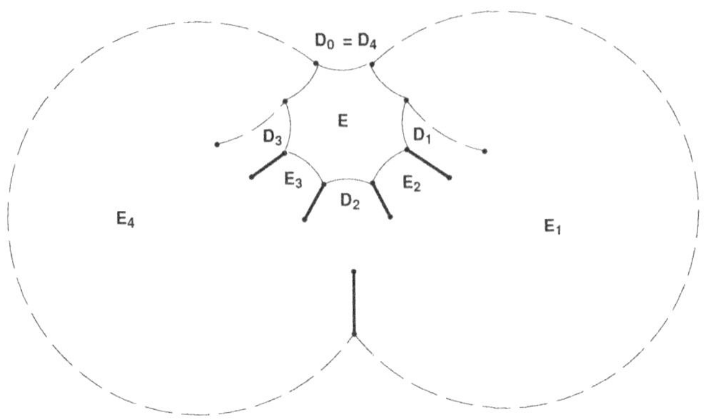


**Figure 8m3.**

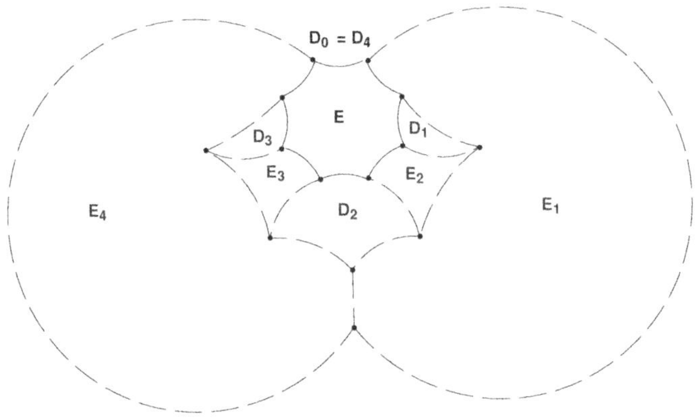


**Figure 8m3a.**

directed path goes from $D_1$ into $E_2$. Since we cannot have a circuit the directed edges goes from $E_1$ into $E_2$. Since all the dead spaces meeting an encumbered component have the same signed curvature the directed edge goes from $D_2$ into $E_2$. Thus all the directed paths from neighboring components go into $E_2$ except the directed path with the sink $E$, but the total curvature of that edge is less than $2\pi/3$. Hence the sum of total curvatures of $E_2$ is less than the required $2\pi/3$ for a component with four edges, Lemma 8, a contradiction.

Suppose $e_{1,4}$ ends at $E_3$, Figure 8m3b. Then $D_3$ is separated from $E_1$ and $E_2$, so it meets just $E_4$ and $E_3$ and must have the curvatures shown. The two outer edges of $D_3$ determine the curvature of the third edge at the vertex at which they meet. We do not know whether the edge $e_{1,3}$ separating $E_1$ and $E_3$ ends on $D_2$. $E_2$, or $D_1$. Should there be an additional dead space at the juncture of $E_1$, $E_3$, and $E_4$, taking the union of the dead space with $E_3$ yields the contradiction.

If $e_{1,3}$ ends on $D_2$, Figure 8m3bi, we see the curvatures of $D_2$ with $E_3$ and $E_1$, which determine the curvature of the edge separating $E_1$ and $E_3$. But now the only edge of $E_3$ which has positive total curvature is the edge with $E$ which contributes too little for optimality.

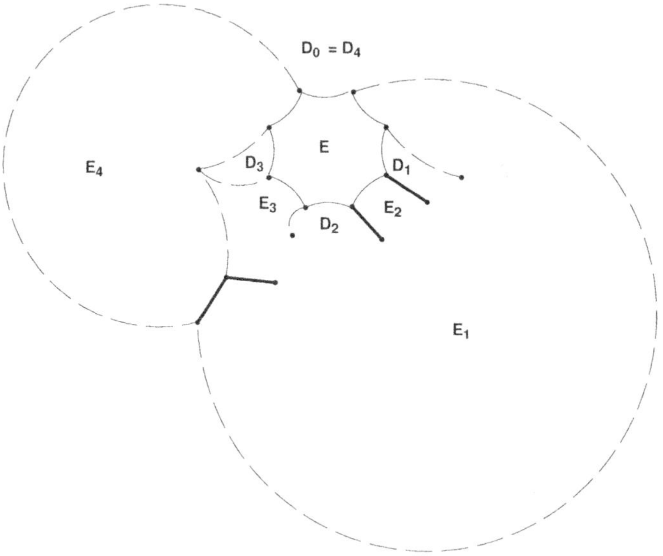


**Figure 8m3b.**

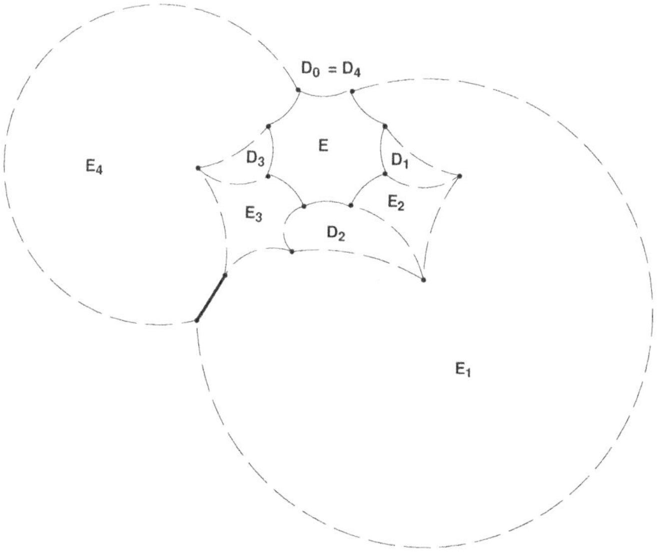


**Figure 8m3bi.**

If $e_{1,3}$ ends on $E_2$, Figure 8m3bii, we deduce that the curvatures of the dead spaces are as shown. Since we can not have a directed path from $D_2$ to $D_0$ we deduce that the edges of $E_3$ are not positive with $E_1$ or $E_4$. Consider $E_1 \cup E_4$. The edges with $D_0$ contribute more than $3\pi$ to the sum of the total curvatures of the edges, while there are no negative contribution. Hence the total curvature sum exceeds the $\pi/3$ required for optimality by Lemma 8. The possible additional dead space where $E_1$, $E_3$, and $E_4$ meet and/or where $E_1$, $E_2$ and $E_3$ meet yield the same contradiction.

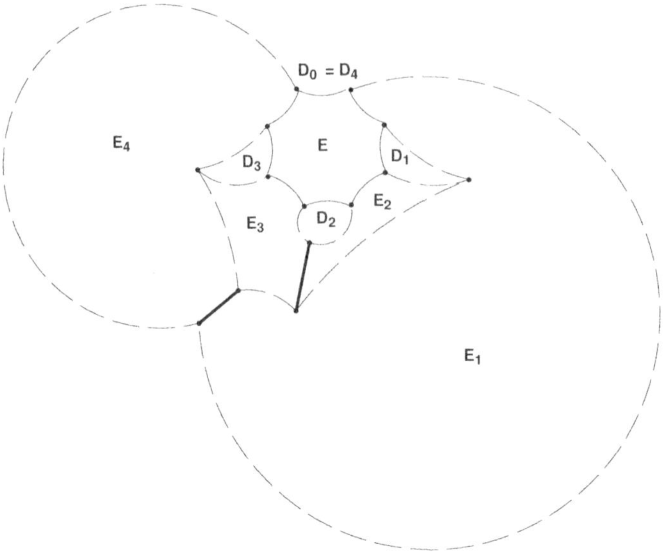


**Figure 8m3bii.**

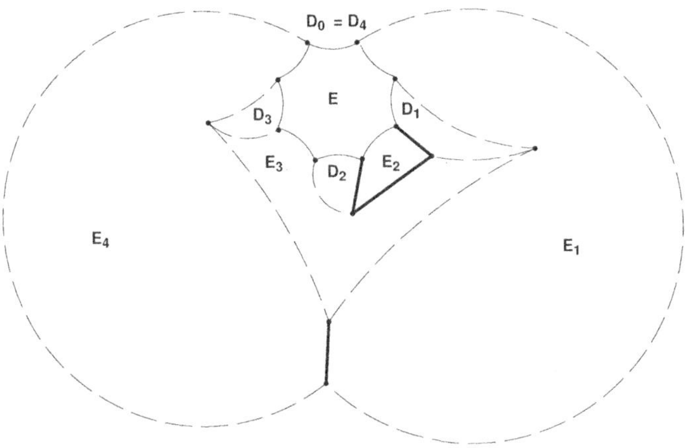


**Figure 8m3biii.**

If $e_{1,3}$ end on $D_1$, then an examination of $D_3$ determines the curvature of $E_3$ with its dead edges as shown in Figure 8m3biii. The vertices of $D_3$ and $D_1$ determine the curvatures of the edges of $E_3$ with both $E_1$ and $E_4$. We now see the that a contradiction will arise by considering $E_1 \cup E_4$, with the usual comment on an additional dead space. Thus $e_{1,4}$ can not end on $E_3$, the Subcase 8m3b is impossible.

If $e_{1,4}$ ends on $D_3$, Figure 8m3c, we do not know where the edge $e_{1,3*}$ separating $E_1$ and $D_3$ ends. Since two dead spaces can not have a common edge, it may end only on $E_3$ or $E_2$.

If $e_{1,3*}$ end on $E_2$ we obtain Figure 8m3ci. Once again examining $E_1 \cup E_4$ yields a contradiction.

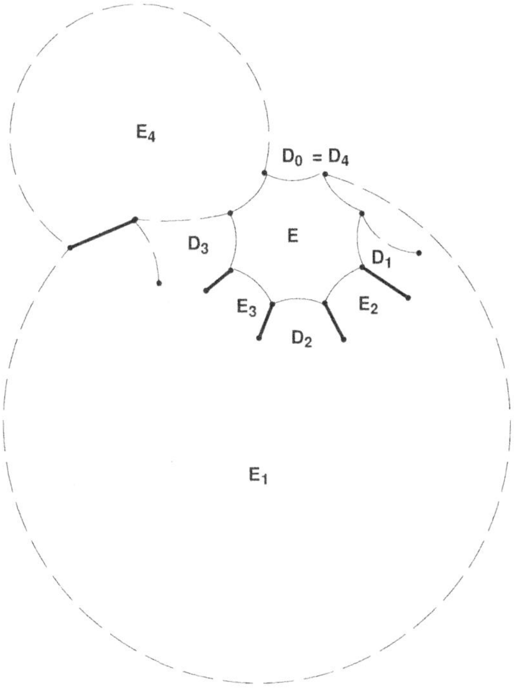


**Figure 8m3c.**

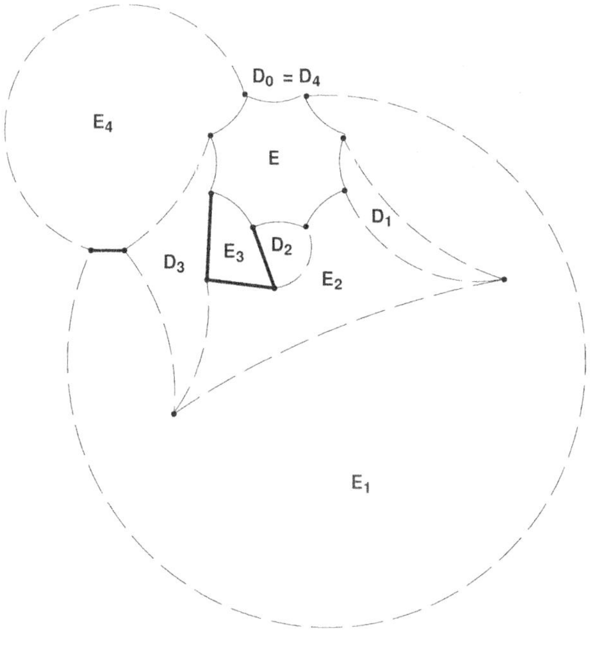


**Figure 8m3ci.**

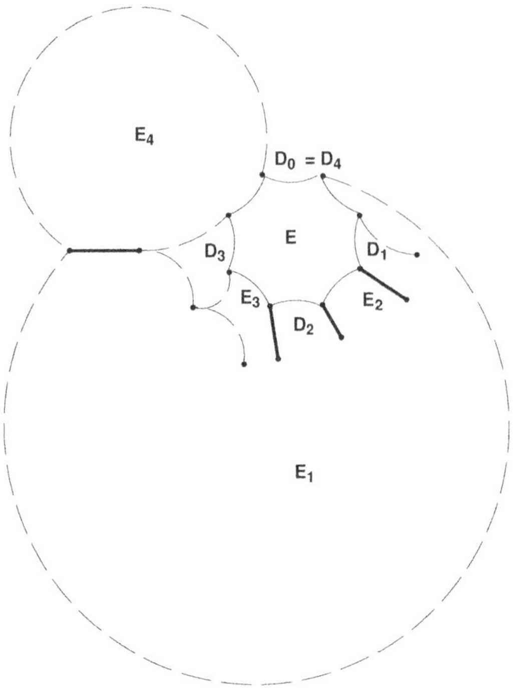


**Figure 8m3cii.**

If $e_{1,3*}$ end on $E_3$, we obtain Figure 8m3cii. Since dead spaces can not meet, $E_2$ must have an additional edge. We do not know where $e_{1,3}$, the edge between $E_1$ and $E_3$, ends. It will end on either $D_2$, $E_2$, or $D_1$.

If $e_{1,3}$ ends on $D_2$, Figure 8m3ciiA, we see that the edge of $E_3$ with $D_3$ has negative curvature for $E_3$, hence it also has negative curvature with $D_2$. Examining the endpoint of $e_{1,3}$ on $D_3$ determines its curvature as shown. Thus the total curvature sum of $E_3$ is insufficient.

If $e_{1,3}$ ends on $E_2$, we obtain Figure 8m3ciiB. In the usual way we note that the total curvature sum of $E_1 \cup E_4$ violates the condition for optimality.

If $E_{1,3}$ end on $D_1$, Figure 8m3ciiC, we again find that $E_1 \cup E_4$ yields a contradiction.

Thus the last possibility for the end of $e_{1,4}$ is also impossible. We conclude that Figure 8m3c can not occur.

We deduce that Case 8m3 can not occur.

**Case 8m4:**

Again, by symmetry, we may suppose $D_0 = D_4$. This means $D_0$ touches $E$, $E_1$, and $E_4$. It must also touch one of $E_2$ or $E_3$. By symmetry we may assume it is $E_2$. See Figure 8m4. Because of $D_1$ not both $E_1$ and $E_2$ can have positive curvatures with $D_0$. So we need to consider the alternatives $E_1$ and $E_4$ have positive curvature with $D_0$ or $E_2$ and $E_4$ do.

Figure 8m4a shows the first alternative. The sign of the curvature of the edges of $E_1$ and $E_2$ with $D_0$ determine the edge of $D_1$, which in turn determine that of the edge

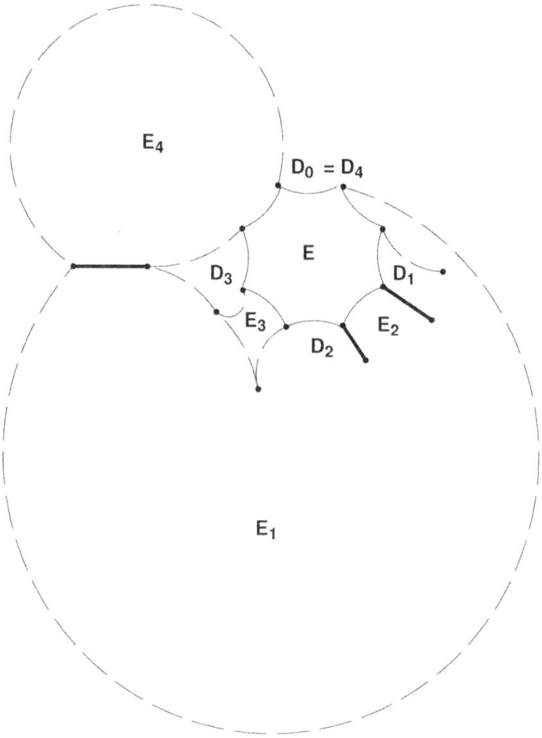


**Figure 8m3ciiA.**

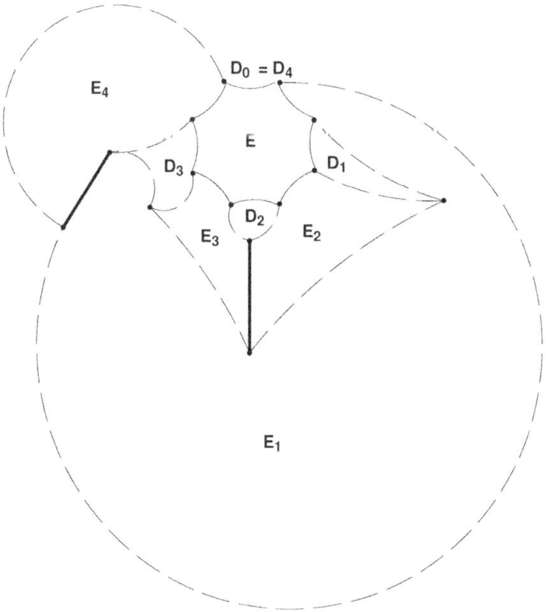


**Figure 8m3ciiB.**

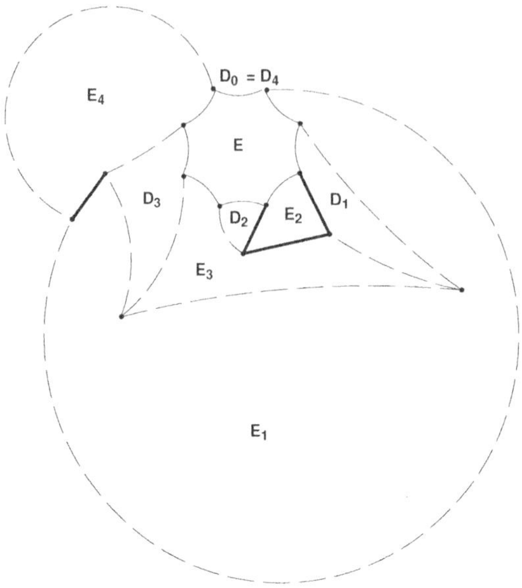


**Figure 8m3ciiC.**

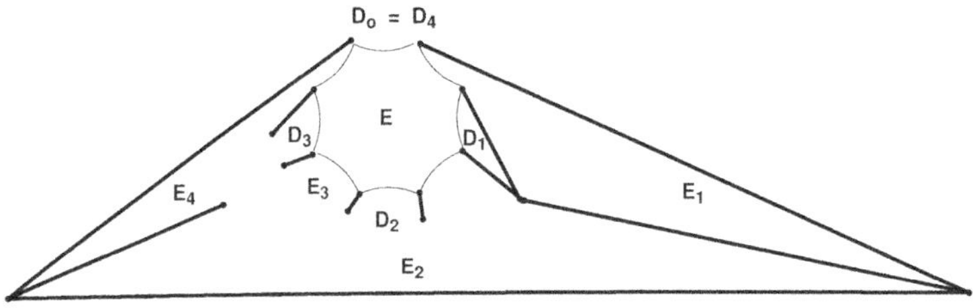


**Figure 8m4.**

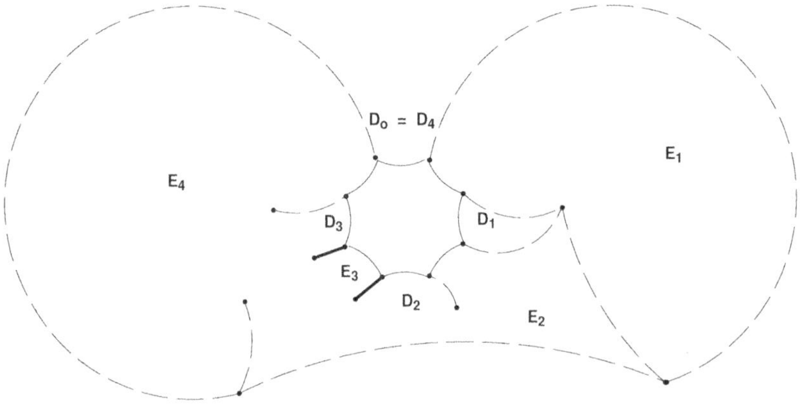


**Figure 8m4a.**

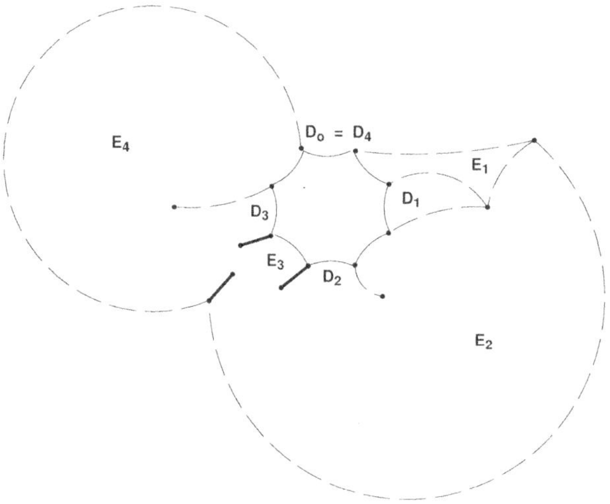


**Figure 8m4b.**

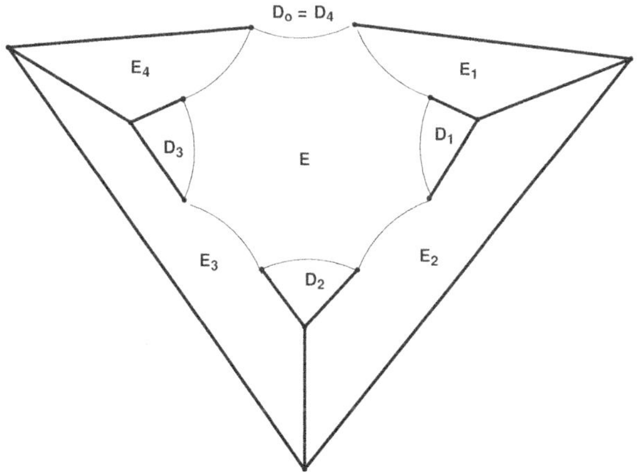


**Figure 8m5.**

separating $E_1$ and $E_2$. But now all the edges of $E_1$ have positive curvature, and the total curvatures of the edge with $D_0$ exceeds $4\pi/3$, which violates optimality by Lemma 8.

The second alternative is illustrated in Figure 8m4b. The edges of $D_1$ and the edge separating $E_1$ and $E_2$ are determined in the same way as the previous alternative, but they all have the opposite sign for their curvatures. The total curvature sum of $E_1$ is insufficient for optimality.

Case 8m4 is impossible.

**Case 8m5:**

In this case all the $E_i$ meet both $E$ and $D_0$. It follows that the dead spaces, $D_i$, $i = 1, 2, 3$, all have three edges. See Figure 8m5.

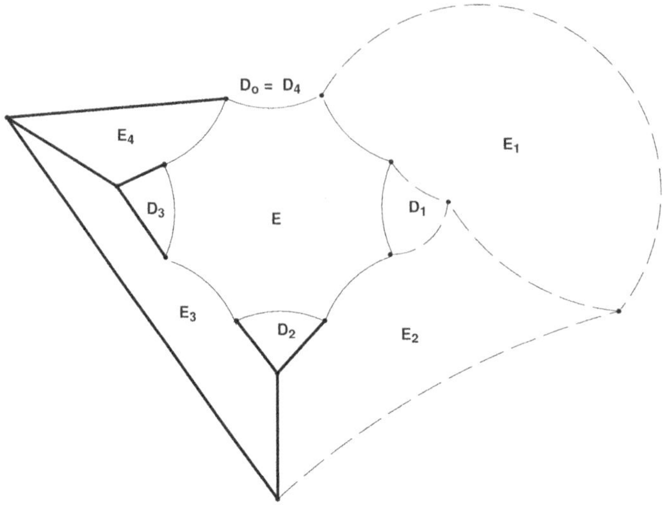


**Figure 8m5a.**

None of the three adjacent pairs $E_1$ and $E_2$, $E_2$ and $E_3$, or $E_3$ and $E_4$ can both have positive curvatures with the dead spaces they touch since, if they did, then the three-edged dead space between them would have positive curvature only on the edge with $E$ which has curvature less than $\pi/3$. Consequently the sum of the total curvatures of that dead space is too small. On the other hand at least two of the $E_i$ must have positive curvatures since the total curvature around $D_0$, viewed from the bounded part, is $2\pi + 5\pi/3$ and the edge with $E$ has negative curvature. We deduce that each edge of $E_i$ with positive curvature with $D_0$ must have total curvature greater than $5\pi/3$ on the edge between them. Thus the following pairs may have positive curvatures with $D_0$ while the others have negative curvatures with $D_0$: $E_1$ and $E_3$, $E_2$ and $E_4$, $E_4$ and $E_1$. The first two possibilities are symmetric, so it suffices to show that it can not be $E_1$ which has positive curvature with $D_0$

If $E_1$ has positive curvature with $D_0$, then it must also have positive curvature with $D_1$. See Figure 8m5a. This forces the other edge of $D_1$ to have positive curvature from the perspective of $D_1$, which forces the curvature the edge between $E_1$ and $E_2$ to be positive from the perspective of $E_1$. But this means each edge of the four-sided region $E_1$ has positive curvature. The sum of the total curvatures of the edges of $E_1$ is greater than that of the edge with $D_0$, which is too large for a four-sided region in an optimal network.

We see that Case 8m5 can not occur.

**Case 8s3:**

In this case $D_0$ does not meet $E$ and $e(D_0) = 3$. Exactly one of the four encumbered regions meeting $E$ does not meet $D_0$. Because of the symmetry we may suppose it is $E_4$. At least two of those meeting $D_0$ have positive curvatures when meeting $D_0$. Because $E_2$ has three-edged dead space at both side, if it has positive curvature neither $E_1$ nor $E_3$ can. Hence we are left with the only possibility that $E_1$ and $E_3$ have positive

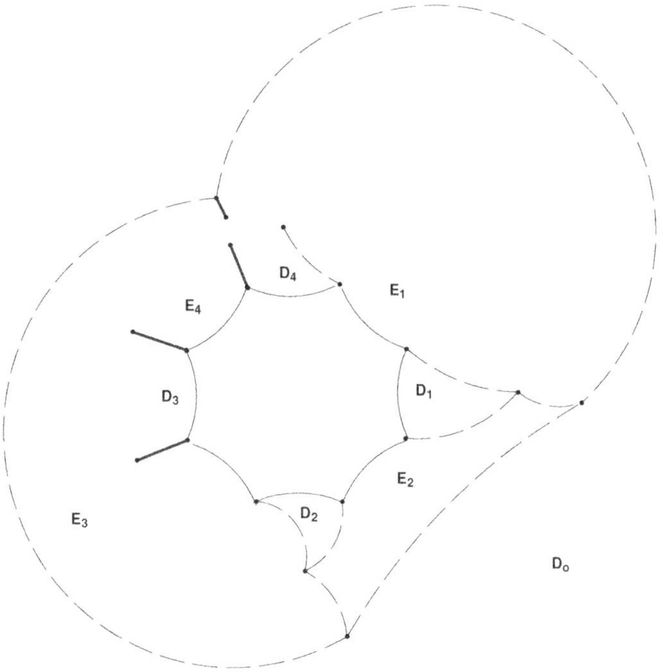


**Figure 8s3.**

curvatures with $D_0$ while $E_2$ does not, Figure 8s3. In this instance the edges of $E_2$ with the dead spaces and with $E_1$ and $E_3$ have negative curvature; the only positive curvature edge of $E_2$ is with $E$. We obtain the following:

$$\sum_{D_1} \tau + \sum_{D_2} \tau + \sum_{E_2} \tau = \pi + \pi + 0,$$

where we have used Lemma 8 to obtain the right-hand side. On the other hand the left-hand side is just the sum of the total curvatures of outer edges of $D_1 \cup D_2 \cup E_1$ all of which are negative except those with $E$ which together contribute less than $2\pi/3$. Hence we reach a contradiction.

Case 8s3 can not occur.

**Case 8s4:**

In this case all the $E_i$ meet $D_0$ and all the dead spaces meeting $E$ have exactly three edges. See Figure 8s4. Consequently no two adjacent $E_i$ can have positive curvatures with $D_0$, but at least two must. By symmetry we may suppose that $E_1$ and $E_3$ are the two with positive curvatures with $D_0$. From the fact that all dead spaces meeting a fixed $E_i$ have the same curvature, we deduce the curvature of the edges of the $D_i$. See Figure 8s4a. The four dead spaces are symmetric and the curvature sign of the edges follows for those of the $E_i$. The vertices of the $D_i$ with the $E_i$ determine the sign of the curvatures of the edges between the $E_i$. We see that $E_1$ has only edges with positive total curvatures, but as a six-sided component, it is only optimal if the sum of the total curvatures would be zero.

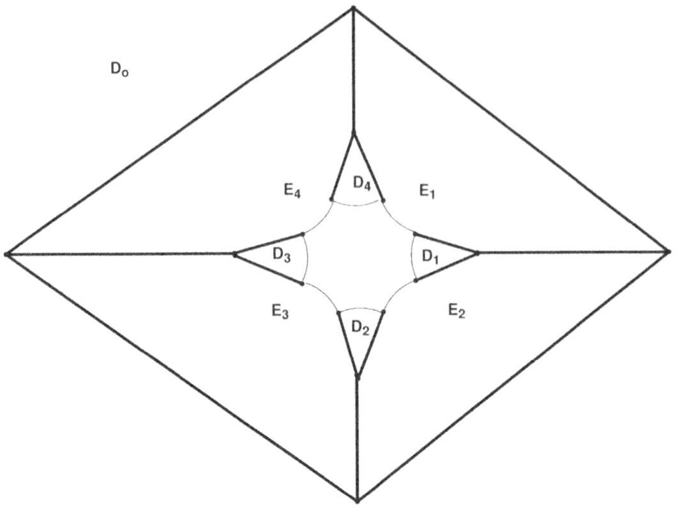


**Figure 8s4.**

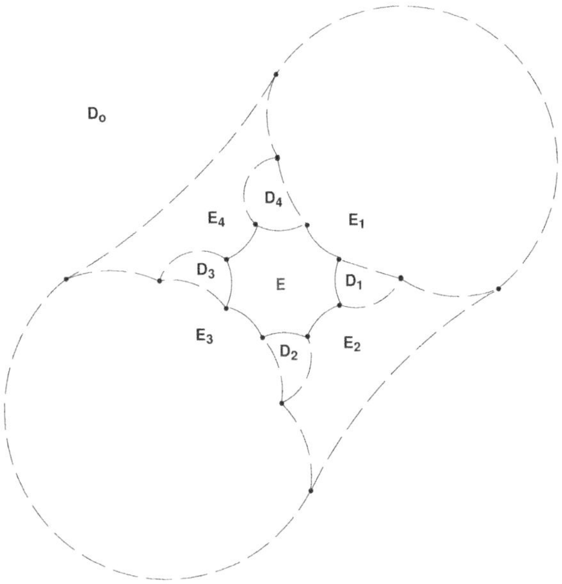


**Figure 8s4a.**

Case 8s4 can not occur.

Since none of these cases can occur, it is impossible that a encumbered component be a sink. Theorem 5 is proved. □

**Corollary 1.** The only sink is $D_0$

*Proof.* There must be a sink, it can be nether an encumbered region nor a dead space. □

**Corollary 2.** The only free space is $D_0$.

*Proof.* If there where a dead space it would have at most five edges, so at least one edge would have positive curvature. The path beginning with the directed edge across that edge with positive curvature would eventually end at the only sink, $D_0$. But this would yield a path beginning and ending at a free space, which is impossible in an optimal configuration, Lemma 14. □

Since the only free space is $D_0$, the main theorem, Theorem 2, is proved.

It may be possible to push this method a little farther, but without a new idea the number of cases would proliferate.[1]

## References


[1] F. J. Almgren, Jr., Existence and regularity almost everywhere of solutions to elliptic variational problems with constraints, *Mem. AMS*, **35** (1976) Providence.

[2] M. N. Bleicher, Isoperimetric division into a finite number of cells in the plane, *Studia Sci. Math. Hung.*, **22** (1987).

[3] M. N. Bleicher, Isoperimetric division into several cells with natural boundary, *Colloq. Math.*, **48** (1985).

[4] C. Cox, L. Harrison, M. Hutchings, S. Kim, J. Light, A. Mauer and M. Tilton, The shortest enclosure of three connected areas in the $\mathbf{R}^2$, Real Analysis Exchange 20, (1994/95).

[5] H. G. Eggleston, Convexity Camb. Tracts in Math. & Math. Phys. #47, Camb. U. P., Cambridge, 1958.

[6] G. Fejes Tóth, An isoperimetric problem for tessellations, *Studia Math. (Hung.)*, **10** (1975).

[7] L. Fejes Tóth, On the shortest nets with meshes of equal area, *Acta Math. Acad. Sci. Hung.*, **11** (1960).

[8] L. Fejes Tóth, Isoperimetric problems concerning tessellations, *Acta Math. Acad. Sci. Hung.*, **14** (1963).

[9] L. Fejes Tóth, What the bees know and what they do not know, *Bull. AMS*, **70** (1964).

[10] L. Fejes Tóth, Regular Figures Int'l Ser. Mono. on Pure and Ap. Math., **48,** Macmillan, New York, 1964.

[11] L. Fejes Tóth, Lagerungen in der Eben auf der Kugel und im Raum (zweite Auflage) Springer-Verlag, Heidelberg, 1972.

[12] L. Fejes Tóth, Isoperimetric problems for tilings, Mathematika, **32** (1985).

[13] J. Foisy, M. Alfaro, J. Brock, N. Hodges and J. Zimba, The standard soap bubble in $R^2$ uniquely minimizes perimeter, *Pac. J. of M.*, **159** (1993).

[14] B. Grünbaum, Convex Polytopes, Interscience Publishers, John Wiley & Sons, New York, 1967.

[15] J. Hass and F. Morgan, Geodesics and soap bubbles in surfaces, *Math. Z.*, **223** (1996).

[16] F. Morgan, Soap bubbles in $R^2$ and in surfaces, *Pac. J. Math.*, **165** (1994).

[17] G. Pólya Induction and analogy in mathematics. Mathematics and plausible reasoning 1, Princeton U. P., Princeton, 1954.

[18] G. Pólya and G. Szegϕ, Isoperimetric inequalities in mathematical physics, *Annal of Math. Studies* **27** Princeton U. P., Princeton, 1951.

[19] M Ritoré and A. Ros, Recent advances in isoperimetric problems, Some updates on isoperimetric problems, *Mathematical Intelligencer*, **24** (2002) 9–14.

[20] A. Siegel, An historical review of the Isoperimetric Theorem in 2-*D* and its place in elementary plane geometry, http://www.cs.nyu.edu/faculty/siegel/SCIAM.pdf.


[1] The authors wish to thank the referees for their helpful comments.

[21] Y. Tomonaga, Geometry of length and area 1, *Dept. of Math.*, Utsunomiya U. Utsunomiya, 1974.

[22] D. Weaire and N. Rivier, Soap cells and statics – Random patterns in two dimensions, *Contemp. Phys.*, **25** (1984).

[23] W. Wichiramala, Proof of the planar triple bubble conjecture, *J. Reine u. Angew. Math.*, **567** (2004).

Proc. Int. Conf. – Number Theory and Discrete Geometry No. 6, 2008, pp. 119–127.

# View-Obstruction Problems

R. J. Hans-Gill and V. C. Dumir*

Department of Mathematics, Panjab University, Chandigarh 160 014, India
e-mail: hansgill@pu.ac.in

*Dedicated to Professor R. P. Bambah on his 80th birthday*

## 1. Introduction

In the thirteenth chapter entitled 'Visibility' of their book 'Lattice points', P. Erdős, P. M. Gruber and J. Hammer write that "Among the most appealing results of combinatorial geometry are visibility theorems ... ". In the next subsection they proceed to describe the view obstruction problem which the authors have been interested in during the last two decades.Though the story starts with Polya's (1918) orchard problem,where the view is to be obstructed from the origin by trees growing at integral points in a bounded region,we are interested in a slightly different point of view. The problem was first formulated as a number theoretical problem.

J. M. Wills (1968) introduced a number of constants concerning the simultaneous approximation of real numbers by rationals. He studied these constants and made some conjectures. One of the constants introduced by him was

$$\kappa(n) = \inf_{\alpha_i,\ 1\le i\le n} \sup_{q} \min_{1\le i\le n} \|q\alpha_i\|,$$

where $\alpha_1, \alpha_2, \ldots, \alpha_n$ are irrational numbers and $q$ varies over all integers.

Using Kronecker's theorem, he obtained the following equivalent representation

$$\kappa(n) = \inf \max_{0\le x\le 1} \min_{1\le i\le n} \|w_i x\|,$$

where the infimum is taken over all positive integers $w_i$, $1 \le i \le n$.

He proved that

$$\kappa(1) = \frac{1}{2},\ \kappa(2) = \frac{1}{3} \quad \text{and}$$

$$\frac{1}{2n} \le \kappa(n) \le \frac{1}{n+1} \quad \text{for all } n \ge 3 \tag{1}$$

and conjectured the value of $\kappa(n)$.

*expired February 2006

## Wills' Conjecture

$$\kappa(\mathbf{n}) = \frac{\mathbf{1}}{\mathbf{n+1}} \tag{2}$$

Betke and Wills (1972) proved that Wills' conjecture would follow if one could prove: For any natural numbers $k_1, k_2, \ldots, k_n$ with $k_i < k_n = c, 1 \leq i \leq n-1$, $(k_1, \ldots, k_n) = 1$, there exists an integer $g$ satisfying

$$c \leq k_i[1 + (n+1)g] \leq nc \mod (n+1)c, \quad \text{for } 1 \leq i \leq n-1.$$

Using this, they showed that determination of $\kappa(2)$ becomes simple. They were also able to deduce $\kappa(3) = \frac{1}{4}$.

T. W. Cusick (1973) used the identity

$$\|\alpha\| + \left\| \frac{1}{2} - \alpha \right\| = \frac{1}{2}$$

to show that $\kappa(n) = \frac{1}{2} - \frac{1}{2}\lambda(n)$, where

$$\lambda(n) = \sup \min_{0 \leq x_i \leq 1} \max_{1 \leq i \leq n} \left\| w_i x - \frac{1}{2} \right\|,$$

the supremum being taken over all n-tuples $w_1, w_2, \ldots w_n$ of natural numbers.

Cusick gave a geometrical interpretation of $\lambda(n)$, generalized the problem and called it view-obstruction problem. Now we proceed to describe this.

Let $R^n$ be the $n-$ dimensional Euclidean space, $Z^n$ be the integer lattice and $\Lambda$ the shifted lattice $Z^n + \left(\frac{1}{2}, \frac{1}{2}, \ldots, \frac{1}{2}\right)$. Let $\mathcal{C}$ be a closed convex body in $R^n$ containing the origin in its interior. For $\alpha \geq 0$, let $\alpha\mathcal{C} = [\alpha X : X \in \mathcal{C}]$ be the magnification of $\mathcal{C}$ by the factor $\alpha$.

Let $P = \{X = (x_1, x_2, \ldots, x_n) : x_i > 0\}$ denote the positive cone. Define $K(\mathcal{C}) = \inf\{\alpha > 0 :$ every ray starting from the origin and lying in $P$ meets $\Delta(\mathcal{C}, \alpha)\}$.

For two special cases, the box $\mathcal{C} : |x_i| \leq \frac{1}{2}, 1 \leq i \leq n$, of side length 1, and the sphere $B :| X |\leq \frac{1}{2}$ of diameter 1, he introduced the notation $K(\mathcal{C}) = \lambda(n)$ and $K(B) = \nu(n)$ respectively. Since $\lambda(n) = 1 - 2\kappa(n)$, Wills' conjecture is equivalent to the following conjecture of Cusick

$$\lambda(n) = \frac{n-1}{n+1}.$$

Cusick (1973) gave geometrical proofs for the values of $\lambda(2)$, $\lambda(3)$ and $\nu(2)$.

In 1974, Cusick gave different proofs for values of $\lambda(2)$ and $\lambda(3)$ using a method which is partly analytic and partly combinatorial. He also made a conjecture for $\nu(n)$. Several developments have taken place in the last few decades since these conjectures were made. We try to summarize some of these.

## 2. Conjecture for Boxes

Here, we give the history of the results concerning the Wills' Conjecture $\kappa(n) = \frac{1}{n+1}$, which is equivalent to $\lambda(n) = \frac{n-1}{n+1}$.

$$\kappa(\mathbf{2}) = \mathbf{1/3}$$

Wills(1968), Betke and Wills (1972), Cusick (1973, 1974, 1982), Chen (1991), Dumir and Hans-Gill (1991) in [22], Bienia et al. (1998), Renault (2004).

$$\kappa(\mathbf{3}) = \mathbf{1/4}$$

Betke and Wills (1972), Cusick (1973, 1974, 1982), Chen (1991) in [6], Dumir and Hans-Gill (1991) in [22] and [23], Bienia et al. (1998), Renault (2004).

$$\kappa(\mathbf{4}) = \mathbf{1/5}$$

Cusick and Pomerance (1984), Chen (1991), Bienia et al. (1998), Bohman et al. (2001), Renault (2004).

$$\kappa(\mathbf{5}) = \mathbf{1/6}$$

Bohman, Holzman and Kleitman (2001), Renault (2004)*.

Bohman et al. also remark that the technique of their proof for $n = 5$ can also be used to prove the lower dimensional results. It may be remarked that several authors have tried to give different proofs in dimensions 2, 3 in order to find a proof which would lead to a method which can work in higher dimensions. It appears that for proofs in dimensions up to 5, initially some new ideas and lots of computations have been needed. Though Bienia et al. (1998) and Renault (2004) greatly simplified the proofs in dimensions $n = 4$ and 5 respectively, still one will probably have to probe a lot deeper to get the right ideas for a proof in higher dimensions.

Wills (1968) using a simple argument, involving an inequality on lengths of intervals covering the interval [0, 1], showed that

$$\kappa(n) \geq \frac{1}{2n} \quad \text{for all natural numbers } n.$$

Chen(1994) improved this to

$$\kappa(n) \geq \frac{1}{2n - 1 - \frac{1}{2n-3}} \quad \text{for } n \geq 5.$$

This was further improved by Chen and Cusick (1999) who obtained

$$\kappa(n) \geq \frac{a}{p}, \quad \text{for } n \geq 4,$$

where $a$ is any positive integer and $p$ is a prime with

$$p \geq \max\{2a(n-1) - 1, 2(a-1)n + 2\}.$$

*J. Barajas and O. Serra have recently proved the conjecture for $n = 6$.

In particular, for $n \geq 4$

$$\kappa(n) \geq \frac{1}{2n-3} \quad \text{if } 2n-3 \text{ is prime}$$

and

$$\kappa(n) \geq \frac{1}{4n-5} \quad \text{if } 4n-5 \text{ is prime.}$$

The method used here is due to Bienia et al. (1998) which was also used by Cusick (1999) to obtain the special case $\kappa(5) \geq \frac{1}{7}$.

Several authors (e.g. Chen (1991), Dumir and Hans-Gill (1991) in [22]) have observed that in the case $n = 2$, the minimal blocking constants can be determined for each ray through the origin into the positive cone. More precisely, if we consider the ray $OW$, where $W = (w, w')$, $w > 0$, $w' > 0$, then for the box $\mathcal{C} : | x_i | \leq \frac{1}{2}$, $i = 1, 2$, $\inf\{\alpha > 0 : \text{ray OW meets } \Delta(\mathcal{C}, \alpha)\}$ equals

$$\begin{cases} \frac{1}{a+b} & \text{if } \frac{w}{w'} = \frac{a}{b},\ (a, b) = 1 \quad \text{and } a, b \text{ are positive integers of opposite parity.} \\ 0 & \text{otherwise} \end{cases}$$

This gives the sequence of isolations as $1/3, 1/5, 1/7, \ldots$ having limit 0. Evidently 0 is attained for infinitely many rays. Dumir and Hans-Gill [22] used first isolation in this to obtain $\lambda(3)$. Also the result was used in [23] to obtain a Markoff type chain of isolations for the three dimensional box (see also Chen (1993)).

The approximation problem formulated by Wills, which has been called view obstruction problem by Cusick, has been further renamed as Lonely Runner Conjecture by Goddyn in 1994 (see Bienia et al. (1998)). Because of (1.1), to prove Wills' conjecture it is enough to prove $\kappa(n) \geq \frac{1}{n+1}$. This would follow if one can prove that if $w_1, w_2, \ldots, w_n$ are any positive integers then there exists an $x \in [0, 1]$ such that $\|w_i x\| \geq \frac{1}{n+1}$ for $1 \leq i \leq n$. Now let us consider the situation where $n + 1$ persons are running with distinct integral speed $w_i$ around a circular track of unit length. Initially, they start at the same time from the same point called the origin. A runner is called *lonely* at time $t$ if its distance from other runners is at least $\frac{1}{n+1}$. Thus the $i^{th}$ runner is *lonely* at time $t$ if

$$\|(w_i - w_j)t\| = \|w_i t - w_j t\| \geq \frac{1}{n+1} \quad \text{for all } j \neq i.$$

If we suppose that $w_{n+1} = 0$ i.e. the $(n+1)^{th}$ runner is stationary, then saying that $(n+1)^{th}$ runner is lonely is equivalent to

$$\|w_i t\| \geq \frac{1}{n+1} \quad \text{for} \quad 1 \leq i \leq n.$$

Thus we have the following equivalent formulation of Will's conjecture:

## Lonely Runner Conjecture: Every runner gets lonely

Bienia et al. (1998) gave a simple proof for $n = 4$ and used it to prove the following result in graph theory:

Let G be an undirected graph. If G has a nowhere zero flow with at most k different values, then it also has one with values from the set $\{1, 2, \ldots, k\}$.

They use the value of $\kappa(n)$ for $n \leq 4$, to deduce the above results for $k \leq 4$ (For $k \geq 5$, it is implied by the six flow theorem of Seymour (1981)).

## 3. The Problem for Spheres

As we mentioned earlier Cusick (1973) made a conjecture for spheres analogous to that for boxes. We introduce some notation which will also be useful later.

Let $L$ be a ray through the origin lying in the positive cone $P$. Define

$$\nu(\mathcal{C}, L) = \inf\{\alpha : L \text{ meets } \Delta(\mathcal{C}, \alpha)\}.$$

Cusick conjectured that $\nu(n) = \nu(B, L_0)$, where $L_0$ is the ray joining the origin to $(1, 2, \ldots, n)$ lying in $P$ and proved it for $n = 2$. In 1986, Dumir and Hans-Gill proved this conjecture for $n = 3$ by showing that $\nu(3) = \sqrt{\frac{3}{7}}$. They also obtained a Markoff type isolation result, giving an infinite sequence of isolated extreme values converging to $\frac{1}{\sqrt{3}}$. (See also Chen (1991, 1994)). Because of the symmetries of $B$, we can suppose that a ray $L$ in $P$ has a point $W = (1, w_2, w_3)$, $0 < w_2 \leq w_3 \leq 1$ on it. Then

$$\nu(B, L) < \frac{1}{\sqrt{3}} \text{ if } w_2 + w_3 \neq 1 \quad \text{and } \nu(B, L) = \frac{1}{\sqrt{3}} \text{ if } w_2 + w_3 = 1$$

with $w_2$ irrational.

In the remaining cases where $w_2 + w_3 = 1$, exact value of $\nu(B, L)$ is obtained. This is at least $\frac{1}{\sqrt{3}}$ in each case.

In 1986, L. Danzer gave examples to show that Cusick's Conjecture is false for $n = 4, 5$ and for some other small values of $n$. Dumir, Hans-Gill and Wilker (1994) and Chen (1994) independently proved that $\nu(4) = \sqrt{\frac{11}{15}}$. We also obtained a Markoff type chain of isolated extreme values. In (1996), we showed that $\nu(5) = \sqrt{\frac{41}{42}}$. Vineet Kahlon (1995) showed that $\nu(B, L_0) = \sqrt{\frac{n(n-1)}{2(2n+1)}}$ for all $n$. Later the authors, L. Danzer and Madhu Raka verified that Cusick's Conjecture is false for $n \leq 32$ and for several larger values of $n$ upto 100. The main problem seems to be the determination, for all $n$, of $\nu(B, L_n)$ for special rays $L_n$. For example the authors feel that the ray $L_n$ passing through $(1, 1, 2, 3, \ldots, n-1)$ would provide a counter example for all $n \geq 6$. So far Cusick's Conjecture has not been revised i.e. no critical lines have been guessed for general $n$.

Chen (1995, 1996) showed how quantitative forms of Kronecker's theorem can be used to investigate the view-obstruction problems.

## 4. Other convex bodies

As in the case of squares and circles, it is easy to solve the problem in the 2-dimensional case. Dumir, Hans-Gill and Wilker (1993) obtained $K(\mathcal{C})$ for all convex domains in $R^2$ which are symmetric in O and the co-ordinate axes. In fact, $\nu(\mathcal{C}, L)$ is obtained in terms of the polar of the gauge function defining $\mathcal{C}$. They also obtained this constant for the octahedron $| x | + | y | + | z | \leq \frac{1}{2}$. In another paper in the same year they obtained the view obstruction constants $K(\mathcal{C})$ for the $l_p-$ balls, $1 < p < \infty$, in $R^3$. They also proved that if

$$\mathcal{C}_p^n :| x_1 |^p + | x_2 |^p + \cdots + | x_p |^p \leq 1/2^p$$

then $K(\mathcal{C}_p^n)$ decreases in $p$ and increases in $n$ while $\frac{K(\mathcal{C}_p^n)}{n^{\frac{1}{p}}}$ increases in $p$. Probably it also increases in $n$. It satisfies

$$\frac{K(\mathcal{C}_p^n)}{n^{\frac{1}{p}}} \leq w_p \leq (1+p)^{\frac{-1}{p}},$$

where

$$w_p = \sup_n \frac{K(\mathcal{C}_p^n)}{n^{\frac{1}{p}}} = \lim_{n \longrightarrow \infty} \frac{K(\mathcal{C}_p^n)}{n^{\frac{1}{p}}}.$$

Determination of $w_p$ is still open. Chen (1994) obtained $K(\mathcal{C})$ for ellipsoids

$$ax^2 + y^2 + z^2 \leq 1.$$

Chen and Mukhopadhya (2002) obtained $K(\mathcal{C})$ for the regular $n$- gons centered at O using elementary geometry.

## 5. Generalizations

Dumir, Hans-Gill and Wilker (1993, 1996) formulated and studied a more general problem in which rays are replaced by subspaces or their translates. One of the reasons for doing this was to use the results for higher dimensional subspaces to obtain the classical view-obstruction constant. In these papers, some basic theory of view-obstruction was developed. This was used by the authors to give a simpler proof of $\nu(3)$ and its isolations in 1996, the determination of $\nu(4)$ and some related isolation results and that of $\nu(5)$ as mention in Section 3. We first give some notations.

Let $\mathcal{C}$ be a closed, centrally symmetric convex body centered at the origin. For each subspace L of $R^n$, define

$$\nu(\mathcal{C}, L) = \inf\{\alpha > 0 : L \cap \Delta(\mathcal{C}, \alpha) \neq \phi\}$$

and for each $d$, $0 < d \leq n$

$$\nu(\mathcal{C}, d) = \sup\{\nu(\mathcal{C}, L) : \dim\ L = d, L \cap P \neq \phi\}.$$

If $C$ is symmetric by reflection in the coordinate hyperplanes, this can be written as $\nu(C, d) = \sup\{\nu(C, L) : L$ is not contained in a co-ordinate hyperplane.$\}$

We find this definition more appropriate, even when the above symmetry condition does not hold. (It may be remarked that the concept can be defined and many analogous results can be obtained in case $C$ contains O in its interior but does not have it as center.)

We obtain a general isolation result by showing that for every $k > 0$, there exist finitely many hyperplanes (which depend on $C$) such that if $L$ is a subspace not lying in one of these hyperplanes, then $\nu(C, L) < k$. It is easy to determine $\nu(C, n-1)$. We determined $\nu(B, n-2)$ for the balls. In 2003, we determined $\nu(B, n-3)$ for $n \geq 7$ and conjectured that $\nu(B, n-k) = \frac{k}{\sqrt{3}}$ if $n \geq 2k+1$. We found that it was natural to extend the concept when subspaces are replaced by flats. This leads to a generalisation of Schoenberg's Billiard Ball Problem. We obtained a variety of analogous results. In particular Schoenberg's conjecture was proved for balls for $n \leq 4$. Hans-Gill (2002) also obtained the corresponding result for octahedron, its isolations and for $l_p-$ balls in $R^3$ for $1 < p < \infty$. We remark that analogous and related concepts for general lattices have been studied by Kannan and Lovász (1988) and by Bambah and Woods (1994) and some other authors.

## References


[1] R. P. Bambah, A. C. Woods, On a problem of G. Fejes Tóth. *Proc. Indian Acad. Sci. Math. Sci.* **104** (1994) no. 1 137–156

[2] U. Betke, J. M. Wills, Untere Schranken für zwei diophantische Approximations-Funktionen. (German) *Monatsh. Math.* **76** (1972) 181–198

[3] W. Bienia, L. Goddyn, P. Gvozdjak, Sebö, A. Tarsi, M. Flows, view obstructions, and the lonely runner. *J. Combin. Theory Ser. B* **72** (1998) no. 1 1–9

[4] T. Bohman, R. Holzman, D. Kleitman, Six lonely runners. Electron. *J. Combin.* **8** (2001) no. 2 Research Paper 3, 49 pp. (electronic)

[5] Chen, Yong-Gao, On a conjecture about Diophantine approximations. I. (Chinese) *Acta Math. Sinica* **33** (1990) no. 5 712–717

[6] Chen, Yong-Gao, On a conjecture in Diophantine approximations. II *J. Number Theory* **37** (1991) no. 2 181–198

[7] Chen, Yong-Gao, On a conjecture in Diophantine approximations. III *J. Number Theory* **39** (1991) no. 1 91–103

[8] Chen, Yong-Gao, On a conjecture in Diophantine approximations. IV *J. Number Theory* **43** (1993) no. 2 186–197

[9] Chen, Yong-Gao, The view-obstruction problem for 3-dimensional spheres *Acta Math. Sinica (N.S.)* **10** (1994) no. 2 158–167

[10] Chen, Yong-Gao, The view-obstruction problem for 4-dimensional spheres *Amer. J. Math.* **116** (1994) no. 6 1381–1419

[11] Chen, Yong-Gao, Ray-obstruction problems for three-dimensional ellipsoids. (Chinese) *Acta Math. Sinica* **37** (1994) no. 1 32–42

[12] Chen, Yong-Gao, View-obstruction problems and Kronecker's theorem. *Proc. Amer. Math. Soc.* **123** (1995) no. 11 3279–3284

[13] Chen, Yong-Gao, View-obstruction problems and Kronecker's theorem. II. *Sci. China Ser. A* **39** (1996) no. 4 347–358

[14] Chen, Yong-Gao, T. W. Cusick, The view-obstruction problem for $n$-dimensional cubes. *J. Number Theory* **74** (1999) no. 1 126–133

[15] Chen, Yong-Gao, Mukhopadhyay, Anirban, The view-obstruction problem for polygons. *Publ. Math. Debrecen* **60** (2002) no. 1–2 101–105

[16] T. W. Cusick, View-obstruction problems. *Aequationes Math.* **9** (1973) 165–170

[17] T. W. Cusick, View-obstruction problems in $n$-dimensional geometry. *J. Combinatorial Theory Ser. A* **16** (1974) 1–11

[18] T. W. Cusick, View-obstruction problems. II. *Proc. Amer. Math. Soc.* **84** (1982) no. 1 25–28

[19] T. W. Cusick, C. Pomerance, View-obstruction problems. III. *J. Number Theory* **19** (1984) no. 2 131–139

[20] T. W. Cusick, The view-obstruction problem for 5-dimensional cubes. *Monatsh. Math.* **127** (1999) no. 3 183–187

[21] V. C. Dumir, R. J. Hans-Gill, View-obstruction problem for 3-dimensional spheres. *Monatsh. Math.* **101** (1986) no. 4 279–289

[22] V. C. Dumir, R. J. Hans-Gill, The view obstruction problems for boxes. *J. Indian Math. Soc. (N. S.)* **57** (1991) no. 1–4 117–122

[23] V. C. Dumir, R. J. Hans-Gill, Markoff type chain for the view obstruction problem for three-dimensional cubes. *J. Indian Math. Soc. (N. S.)* **57** (1991) no. 1–4 123–141.

[24] V. C. Dumir, R. J. Hans-Gill, J. B. Wilker, Contributions to a general theory of view-obstruction problems. *Canad. J. Math.* **45** (1993) no. 3 517–536

[25] V. C. Dumir, R. J. Hans-Gill, J. B. Wilker, The $l_p$ family of view-obstruction problems. *Nieuw Arch. Wisk.* (4) **11** (1993) no. 3 225–239

[26] V. C. Dumir, R. J. Hans-Gill, J. B. Wilker, The view-obstruction constants for $l_p$-balls in $R^3$. *Exposition. Math.* **11** (1993) no. 5 407–417

[27] V. C. Dumir, R. J. Hans-Gill, J. B. Wilker, The view-obstruction problem for spheres in $R^5$. *Monatsh. Math.* **122** (1996) no. 1 21–34

[28] V. C. Dumir, R. J. Hans-Gill, J. B. Wilker, Contributions to a general theory of view-obstruction problems. *II. J. Number Theory* **59** (1996) no. 2 352–373

[29] V. C. Dumir, R. J. Hans-Gill, View-obstruction through trajectories of co-dimension three. *Discrete geometry*, 229–237, *Monogr. Textbooks Pure Appl. Math.* **253** Dekker, New York, (2003)

[30] P. Erdös, P. M. Gruber, J. Hammer, Lattice points. Pitman Monographs and Surveys in Pure and Applied Mathematics, 39. Longman Scientific and Technical, Harlow; copublished in the United States with John Wiley and Sons, Inc., New York, (1989). viii+184

[31] R. J. Hans-Gill, The billiard ball motion problem. I. A Markoff type chain for the octahedron in $R^3$. Number theory and discrete mathematics (Chandigarh, 2000), 213–223, *Trends Math.*, Birkhäuser, Basel, (2002)

[32] R. J. Hans-Gill, The billiard ball motion problem. II. $l_p$-balls in $R^3$, $1 < p < \infty$. IV International Conference in "Stochastic Geometry, Convex Bodies, Empirical Measures and Applications to Engineering Science", Vol. I (Tropea, 2001). *Rend. Circ. Mat. Palermo (2) Suppl.* no. 70, part I (2002) 347–360

[33] Kahlon, Vineet, View obstruction problem. *Res. Bull. Panjab Univ. Sci.* **45** (1995) no. 1–4 175–180

[34] R. Kannan, L. Lovász, Covering minima and lattice-point-free convex bodies. *Ann. of Math.* (2) **128** (1988) no. 3 577–602

[35] Mukhopadhyay, Anirban, The view-obstruction problem. Currents trends in number theory (Allahabad, 2000), 91–100, Hindustan Book Agency, New Delhi, (2002)

[36] Renault, Jerome, View-obstruction: a shorter proof for 6 lonely runners. *Discrete Math.* **287** (2004) no. 1–3 93–101

[37] I. J. Schoenberg, Extremum problems for the motions of a billiard ball. I. The $L_p$ norm, $1 \leq p < \infty$. *Nederl. Akad. Wetensch. Proc. Ser. A 79=Indag. Math.* **38** (1976) no. 1 66–75

[38] I. J. Schoenberg, Extremum problems for the motions of a billiard ball. II. The $L_\infty$ norm. *Nederl. Akad. Wetensch. Proc. Ser. A 79=Indag. Math.* **38** (1976) no. 3 263–279

[39] I. J. Schoenberg, Extremum problems for the multi-dimensional case of König and Szücs of billiard ball motions. *C. R. Math. Rep. Acad. Sci. Canada* **1** (1978/79) no. 1 37–40

[40] I. J. Schoenberg, Extremum problems for the motions of a billiard ball. III. The multidimensional case of König and Sücs. *Studia Sci. Math. Hungar.* **13** (1978) no. 1–2 53–78 (1981)

[41] P. D. Seymour, Nowhere-zero 6-flows. *J. Combin. Theory Ser.* **B30** (1981) no. 2 130–135

[42] Wills, M. Jörg, Zur simultanen homogenen diophantischen Approximation. I. (German) *Monatsh. Math.* **72** (1968) 254–263

Proc. Int. Conf. – Number Theory and Discrete Geometry No. 6, 2008, pp. 129–142.

# Minkowski's Successive Minima

Martin Henk[1] and Jörg M. Wills[2]

[1]Martin Henk, Universität Magdeburg, IAG, Universitätsplatz 2, D-39106 Magdeburg, Germany
e-mail: henk@math.uni-magdeburg.de
[2]Jörg M. Wills, Universität Siegen, Mathematisches Institut, ENC, D-57068 Siegen, Germany
e-mail: wills@mathematik.uni-siegen.de

**Abstract.** In this note we give a brief survey on some classical and old problems involving Minkowski's successive minima of a 0-symmetric convex body as well as on some recent developments concerning these functionals. Among others we show that the successive minima are closely related to intrinsic volumes, the lattice point enumerator, the zeros of the Ehrhart polynomial and to simultaneous Diophantine approximation problems.

**2000 Subject Classification:** 11H06, 52C07

*Dedicated to R. P. Bambah on the occasion of his 80th birthday*

## 1. Introduction

One of the basic questions in geometry of numbers is to decide whether a given set in the $n$-dimensional Euclidean space $\mathbb{R}^n$ contains a lattice point of the intgeral lattice $\mathbb{Z}^n$. With respect to the class $\mathcal{K}_0^n$ of all 0-symmetric convex bodies, i.e., convex compact subsets $K \subset \mathbb{R}^n$ with $K = -K$ and $\dim(K) = n$, and the volume functional $\mathrm{vol}(\cdot)$, Minkowski settled this problem [40, pp.75]:

$$\textit{If } \mathrm{vol}(K) \geq 2^n \textit{ then } K \textit{ contains a non-zero lattice point.} \tag{1.1}$$

The $n$-cube $C_n = \{x \in \mathbb{R}^n : |x_i| \leq 1,\ 1 \leq i \leq n\}$ of volume $2^n$ shows that, in general, the constant $2^n$ can not be replaced by a smaller one. Although the proof of (1.1) is rather simple it has a lot of applications in different branches of mathematics (cf., e.g., [27, pp. 40], [40, pp. 102]). Minkowski also proved an important generalisation of (1.1) for which we have to introduce his successive minima. For $K \in \mathcal{K}_0^n$ and $1 \leq i \leq n$ the $i$-th successive minimum $\lambda_i(K)$ is defined by

$$\lambda_i(K) = \min\{\lambda > 0 : \dim(\lambda\, K \cap \mathbb{Z}^n) \geq i\},$$

i.e., $\lambda_i(K)\, K$ is the smallest dilate of $K$ containing $i$ linearly independent lattice points. Obviously, $\lambda_i(K) \leq \lambda_{i+1}(K)$ and $\mathrm{int}(\lambda_1(K)\, K) \cap \mathbb{Z}^n = \{0\}$, where $\mathrm{int}()$ denotes the interior. Moreover, $\lambda_i(K)$ is homogeneous of degree $-1$, which means that $\lambda_i(\mu\, K) = (1/\mu)\, \lambda_i(K)$ for any positive number $\mu$. With this notation, (1.1) can be equivalently reformulated as

**Theorem 1.1 (Minkowski's 1st theorem on successive minima).** *Let $K \in \mathcal{K}_0^n$. Then*

$$\mathrm{vol}(K) \leq \left(\frac{2}{\lambda_1(K)}\right)^n. \tag{1.2}$$

To see that (1.1) and (1.2) are equivalent, we firstly note that for $K \in \mathcal{K}_0^n$ with $\mathrm{vol}(K) \geq 2^n$, (1.2) implies $\lambda_1(K) \leq 1$ and thus $K$ must contain a non-trivial lattice point. On the other hand, since $\mathrm{int}(\lambda_1(K)\,K) \cap \mathbb{Z}^n = \{0\}$, (1.1) shows that $\mathrm{vol}(\lambda_1(K)\,K) \leq 2^n$ and thus we get (1.2). The problem to classify those bodies satisfying (1.2) with equality leads to the theory of extremal bodies (see [27, pp. 82]).

As mentioned above, Minkowski proved a generalisation of (1.2) which is given by the upper bound in the next theorem (cf. [40, pp. 187], [27, pp. 59]).

**Theorem 1.2 (Minkowski's 2nd theorem on successive minima).** *Let $K \in \mathcal{K}_0^n$. Then*

$$\frac{1}{n!}\prod_{i=1}^{n}\frac{2}{\lambda_i(K)} \leq \mathrm{vol}(K) \leq \prod_{i=1}^{n}\frac{2}{\lambda_i(K)}. \tag{1.3}$$

Both bounds are best possible. For instance, the lower bound is attained by the regular $n$-cross polytope $C_n^\star = \{x \in \mathbb{R}^n : \sum_{i=1}^n |x_i| \leq 1\}$. Here we have $\mathrm{vol}(C_n^\star) = 2^n/n!$ and $\lambda_i(C_n^\star) = 1$, $1 \leq i \leq n$. Of course, the upper bound is still tight for the cube $C_n$, but also for arbitrary boxes $Q$ with facets parallel to the coordinate hyperplanes. For instance, let

$$Q = \{x \in \mathbb{R}^n : |x_i| \leq \alpha_i,\ 1 \leq i \leq n\} \tag{1.4}$$

where we may assume $\alpha_i \leq \alpha_{i-1}$. Then $\lambda_i(Q) = 1/\alpha_i$ and thus we have equality in the upper bound. The lower bound is easy to prove. For if, choose $n$-linearly independent lattice points $z_i$ such that $z_i \in \lambda_i(K)\,K$, $1 \leq i \leq n$. Then $K$ contains the cross polytope $\mathrm{conv}\{\pm(1/\lambda_i(K))\,z_i : 1 \leq i \leq n\}$ and the volume of that cross polytope is at least as large as the left hand side in (1.3). In contrast to the lower bound the upper bound is considered as a rather deep result in geometry of numbers. This is also reflected by the fact that many eminent mathematicians, e.g., Bambah, Woods, Zassenhaus, Davenport, Siegel, Weyl, etc., tried to improve on Minkowski's second theorem and gave themselves interesting proofs (cf., e.g., [3,15,50,55]). For a modern version of Minkowski's original proof we refer to [29].

The main purpose of this note is to give a brief survey on classical problems concerning the successive minima of a 0-symmetric convex body and also to show new developments and further relations of the successive minima to other geometric functionals. We will state the results mainly with respect to the standard lattice $\mathbb{Z}^n$. Most of the results, however, can easily be generalised to arbitrary lattices by a linear transformation.

At this point we also want to remark that Minkowski's theory of successive minima has been applied successfully in a much broader context than presented here; we just mention the fields i) adelic geometry of numbers (cf. [54]), ii) projective toric varieties (cf. [51]) and iii) successive minima of indefinite quadratic forms (cf. [4,13,21]).

The paper is organised as follows. In Section 2 we will study some classical generalisation of Minkowski's theorems. Then we want to point out relations between the so called intrinsic volumes and the successive minima. In Section 4 we show that there are also bounds on the number of lattice points contained in a 0-symmetric convex body in terms of the successive minima. For the class of 0-symmetric lattice polytopes there seems to be even some interesting relations between the coefficients of the so called Ehrhart polynomial and the successive minima. This topic is the content of Section 5. Finally, in the last section we show generalisation of Minkowski's fundamental theorems in the context of simultaneous Diophantine approximation problems.

## 2. Generalisations of Minkowski's theorems

The most natural way to extend (1.1) is to ask for a general lower bound on the number of lattice points contained in $K \in \mathcal{K}_0^n$ with respect to the volume. Here we have the following result due to van der Corput [14], [27, pp. 51]

**Theorem 2.1 (van der Corput).** *Let $K \in \mathcal{K}_0^n$. Then*

$$\#(K \cap \mathbb{Z}^n) \geq 2\left\lfloor \frac{\mathrm{vol}(K)}{2^n} \right\rfloor + 1,$$

where $\lfloor x \rfloor$ denotes the integral part of $x \in \mathbb{R}$. In particular, if $\mathrm{vol}(K) \geq 2^n$ we get (1.1). A generalisation of a different type is due to Siegel [49], [27, pp. 50] which is based on Parseval's identity for multiple Fourier series.

**Theorem 2.2 (Siegel).** *Let $K \in \mathcal{K}_0^n$ with $\mathrm{int} K \cap \mathbb{Z}^n = \{0\}$. Then*

$$\mathrm{vol}(K) + \frac{\mathrm{vol}(K)}{4^n} \sum_{z \in \mathbb{Z}^n \setminus \{0\}} \left| \int_{(1/2)K} \mathrm{e}^{-2\pi \mathrm{i}\, z \cdot x}\, \mathrm{d}x \right|^2 = 2^n,$$

where $x \cdot y$ denotes the usual inner product on $\mathbb{R}^n$. Since the second term on the left hand side is non-negative Theorem 2.2 implies (1.1). In general, however, it is difficult to evaluate the integral above and thus it is hard to take advantage of this beautiful identity of Siegel.

In order to present a direct sharpening of (1.2) we need the notation of the density of a densest lattice packing. For details we refer to [27, pp. 218]. A lattice $\Lambda \subset \mathbb{R}^n$ is called a packing lattice of a set $K \in \mathcal{K}_0^n$ if $\mathrm{int}(2K) \cap \Lambda = \{0\}$, i.e., two different lattice translates $a_1 + K, a_2 + K, a_1 \neq a_2 \in \Lambda$, have no interior points in common. The density $\delta(K, \Lambda)$ of such a non-overlapping lattice arrangement $\Lambda + K$ is the proportion of space which is occupied by the translates $\Lambda + K$, and it is given by

$$\delta(K, \Lambda) = \frac{\mathrm{vol}(K)}{\det \Lambda},$$

where $\det \Lambda$ denotes the determinant of the lattice. The maximum density $\delta(K, \Lambda)$ with respect to all packing lattices $\Lambda$ of $K$ is called the density of a densest lattice packing of $K$ and it will be denoted by $\delta(K)$.

By the definition of the first successive minimum we know that $\mathbb{Z}^n$ is a packing lattice of the body $(\lambda_1(K)/2)\,K$. Hence we have $\delta(K) \geq \mathrm{vol}((\lambda_1(K)/2)\,K)$ or

$$\mathrm{vol}(K) \leq \delta(K)\left(\frac{2}{\lambda_1(K)}\right)^n. \tag{2.1}$$

Since $\delta(K) \leq 1$ the inequality above may be considered as an improvement of (1.2), which, in particular, takes more into account the shape of the body $K$. In view of the upper bound in (1.3) it is tempting to improve also (2.1) by replacing $(2/\lambda_1(K))^n$ with the product of the successive minima. This is the content of a famous problem posed by Davenport [16].

**Problem 2.3 (Davenport).** Let $K \in \mathcal{K}_0^n$. Is it true that

$$\mathrm{vol}(K) \leq \delta(K) \prod_{i=1}^{n} \frac{2}{\lambda_i(K)}\ ? \tag{2.2}$$

So far it has only been verified for $n = 2$ and for ellipsoids by Minkowski [40, pp. 196], [27, pp. 195], the case $n = 3$ was settled by Woods [58]. His proof is based on Minkowski's classification of of so called critical lattices in dimension 3 which does not seem to be extendable to higher dimensions (cf. [41], [27, pp. 340]).

For the more general class of ray sets of finite type Rogers and independently Chabauty proved an inequality of type (2.2), but with an additional factor of $2^{(1/2)(n-1)}$ on the right hand side. It was shown, however, by Mahler and Chabauty that in this more general setting the additional factor is necessary (cf. [27, pp. 188]).

There are various attempts to generalise Minkowski's theorem to non-symmetric convex bodies (cf. [27, pp. 52]). Here we just want to mention a still open and quite fascinating conjecture due to Ehrhart [22]

*Conjecture* 2.4 (Ehrhart). Let $K \subset \mathbb{R}^n$ be a convex body whose centre of gravity is the origin. If $\mathrm{vol}(K) \geq (n+1)^n/n!$ then $K$ contains a non-trivial lattice point.

This inequality would be tight as the $n$-dimensional simplex $-\mathbf{1} + (n+1)T_n$ shows, where $\mathbf{1} = \sum_{i=1}^n e_i$ denotes the all 1-vector, $e_i$ the $i$-th unit vector and $T_n = \mathrm{conv}\{0, e_1, \ldots, e_{n+1}\}$ the standard simplex in $\mathbb{R}^n$ of volume $1/n!$. So far the conjecture has only be proven in the plane and for special 3-dimensional convex bodies by Ehrhart [22].

## 3. Intrinsic volumes

Let $B^i$ be the $i$-dimensional unit ball whose $i$-dimensional volume will be denoted by $\kappa_i$. For a convex body $K \subset \mathbb{R}^n$ the outer parallel body at distance $\epsilon > 0$ is given by $K + \epsilon\, B^n$, i.e., it consists of all points whose (Euclidean) distance to $K$ is at most $\epsilon$.

The volume of $K + \epsilon B^n$ is a polynomial of degree $n$ in $\epsilon$, the so called Steiner-polynomial (cf. [48, pp. 197])

$$\mathrm{vol}(K + \epsilon B^n) = \sum_{i=0}^{n} \epsilon^{n-i} \kappa_{n-i} \mathrm{V}_i(K). \tag{3.1}$$

The coefficients $\mathrm{V}_i(K), i = 0, \dots, n$, are the so called intrinsic volumes. They are normalised Minkowskian Quermassintegrals and they were introduced by Peter McMullen [39]. In particular, we have $\mathrm{V}_n(K) = \mathrm{vol}(K)$, $\mathrm{V}_{n-1}(K)$ is one half of the surface area $\mathrm{F}(K)$ and $\mathrm{V}_0(K) = 1$.

Since the volume functional itself is an intrinsic volume and in view of (1.3) one may ask for further inequalities involving the successive minima and other intrinsic volumes. Concerning a lower bound we have the following generalisation of (1.3) [56]

**Theorem 3.1 (W.).** *Let $K \in \mathcal{K}_0^n$. Then for $1 \leq m \leq n$*

$$\frac{1}{m!} \prod_{i=1}^{m} \frac{2}{\lambda_i(K)} \leq \mathrm{V}_m(K).$$

This inequality is tight and for $m = n$ we obtain the lower bound in (1.3). In general, $\lambda_1(K) \cdot \ldots \cdot \lambda_m(K) \cdot \mathrm{V}_m(K)$ can not be bounded from above by a constant. Hence in order to find an extension of the upper bound in (1.3) to other intrinsic volumes a different type of inequality is needed. Here we have [28].

**Theorem 3.2 (H.).** *Let $K \in \mathcal{K}_0^n$. Then for $0 \leq m \leq n - 1$*

$$\mathrm{vol}(K) \leq \mathrm{V}_m(K) \prod_{i=m+1}^{n} \frac{2}{\lambda_i(K)}.$$

Again this inequality is best possible and for $m = 0$ we get Minkowski's upper bound in (1.3). Of particular interest and simplicity is the case $m = n - 1$ which says that

$$\lambda_n(K) \leq \frac{\mathrm{F}(K)}{\mathrm{vol}(K)}. \tag{3.2}$$

So for $\lambda$ bigger than the ratio of surface area to volume the dilate $\lambda\, K$ contains $n$-linearly interior lattice points. Finally we want to remark that it is an open problem to generalise the theorems above to arbitrary lattices (cf. [26]).

## 4. Lattice point enumerator

By definition the successive minima measure or reflect a certain lattice point property of a 0-symmetric convex body. Hence it is quite natural to look for bounds on the lattice point enumerator $\mathrm{G}(K) = \#(K \cap \mathbb{Z}^n)$ in terms of the successive minima. A result in this spirit is again due to Minkowski who proved [40, p. 79]

**Theorem 4.1 (Minkowski).** *Let $K \in \mathcal{K}_0^n$ with $\lambda_1(K) \geq 1$, i.e.,* $\mathrm{int}(K) \cap \mathbb{Z}^n = \{0\}$. *Then*

$$\mathrm{G}(K) \leq 3^n.$$

The cube $C_n$ shows that the bound can not be improved in general. Minkowski also proved a sharper bound of $2^{n+1} - 1$ for the class of strictly 0-symmetric convex bodies, but for simplification we will deal only with the general case and refer to [27, p. 63] for details. In [11] the above result was embedded in a more general inequality, namely

**Theorem 4.2 (Betke,H.,W.).** *Let $K \in \mathcal{K}_0^n$. Then*

$$\mathrm{G}(K) \leq \left\lfloor \frac{2}{\lambda_1(K)} + 1 \right\rfloor^n.$$

This inequality may be considered as a lattice point analogue of Minkowski's first Theorem 1.1. Roughly speaking, we have only replaced the volume by the lattice point enumerator. Moreover, since $K$ is a Jordan-measurable set we have

$$\lim_{m \to \infty} \frac{\mathrm{G}(m\,K)}{\mathrm{vol}(m\,K)} = 1, \tag{4.1}$$

and thus Theorem 4.2 implies Minkowski's first theorem. By the same reasoning the following lower bound proved in [11] is a generalisation of the lower bound in Minkowski's second Theorem 1.2

$$\left(1 - \frac{\lambda_1(K)}{2}\right) \frac{1}{n!} \prod_{i=1}^{n} \frac{2}{\lambda_i(K)} \leq \mathrm{G}(K).$$

Here we have additionally to assume that $\lambda_1(K) \leq 1$.

Of course, the most challenging problem is to give an analogue of the upper bound in Minkowski's second theorem but for that problem, unfortunately, we only have a conjecture [11]

*Conjecture* 4.3 (Betke, H., W.). Let $K \in \mathcal{K}_0^n$. Then

$$\mathrm{G}(K) \leq \prod_{i=1}^{n} \left\lfloor \frac{2}{\lambda_i(K)} + 1 \right\rfloor.$$

As in Minkowski's upper bound we have equality for boxes with facets parallel to the coordinate axes (cf. (1.4)). So far this conjecture has only been verified in the 2-dimensional case [11] and in the general case it is only known that [29]

$$\mathrm{G}(K) \leq 2^{n-1} \prod_{i=1}^{n} \left\lfloor \frac{2}{\lambda_i(K)} + 1 \right\rfloor.$$

Since the conjecture would imply Minkowski's upper bound in (1.3) it seems to be a rather hard problem.

Now we relax a little bit the right hand side in the conjecture and set

$$\mathrm{L}(K) = \prod_{i=1}^{n} \left( \frac{2}{\lambda_i(K)} + 1 \right).$$

Since the successive minima are homogeneous of degree $-1$ we see that for any positive number $\mu$

$$\mathrm{L}(\mu\, K) = \prod_{i=1}^{n} \left( \frac{2}{\lambda_i(K)}\, \mu + 1 \right) \tag{4.2}$$

is a polynomial of degree $n$ in $\mu$ and Conjecture 4.3 relates this polynomial with the lattice point enumerator of a 0-symmetric convex body. There is another and quite famous polynomial related to the number of lattice points which will be discussed in the next section.

## 5. Ehrhart polynomials

In the following we will study the number of lattice points contained in lattice polytopes, which in our setting are polytopes whose vertices belong to the integral lattice $\mathbb{Z}^n$. The space of all $n$-dimensional lattice polytopes in $\mathbb{R}^n$ is denoted by $\mathcal{P}^n$. In 1899 G. Pick [44] found the following beautiful identity for lattice polygons

**Theorem 5.1 (Pick).** *Let $P \in \mathcal{P}^2$. Then*

$$\mathrm{G}(P) = \mathrm{vol}(P) + \frac{1}{2}\mathrm{G}(\mathrm{bd}\, P) + 1,$$

where bd() denotes the boundary. The statement is also true for non-selfintersecting (and not necessarily convex) polygons. The proof of this theorem relies heavily, as many other results on lattice points in the plane, on the 2-dimensional fact that the area of any lattice triangle whose vertices are the only lattice points is bounded. In fact, the volume is always $1/2$. In higher dimensions the so called Reeve simplices [46]

$$R_n(m) = \mathrm{conv}\{0, e_1, \ldots, e_{n-1}, m \textstyle\sum_{i=1}^{n} e_i\},\ m \in \mathbb{N}, \tag{5.1}$$

show that the volume of such a simplex may not be bounded at all.

Recently, M. Ram Murty and Nithum Thain [43] gave a nice proof of Pick's theorem via Minkowski's theorem. Since for any integer $k$ the dilate $k P$ is again a lattice polygon Pick's theorem immediately implies

$$\mathrm{G}(k\, P) = \mathrm{vol}(P)\, k^2 + \frac{1}{2}\mathrm{G}(\mathrm{b}d\, P)\, k + 1,$$

$$\mathrm{G}(\mathrm{int}(k\, P)) = \mathrm{vol}(P)\, k^2 - \frac{1}{2}\mathrm{G}(\mathrm{b}d\, P)\, k + 1.$$

In the years 1962–1968 E. Ehrhart [23–25] generalised these identities to all dimensions

**Theorem 5.2 (Ehrhart).** *Let $P \in \mathcal{P}^n$ and $k \in \mathbb{N}$. Then*

$$\mathrm{G}(k\,P) = \sum_{i=0}^{n} \mathrm{G}_i(P)\,k^i,$$

$$\mathrm{G}(\mathrm{int}(k\,P)) = (-1)^n \sum_{i=0}^{n} \mathrm{G}_i(P)\,(-k)^i,$$

*where the coefficients $\mathrm{G}_i(P)$ depends only on $P$.*

The first identity is known as the Ehrhart polynomial and the second one is called Ehrhart's reciprocity law. Two of the $n+1$ coefficients $\mathrm{G}_i(P)$ are obvious, namely, $\mathrm{G}_0(P) = 1$ and $\mathrm{G}_n(P) = \mathrm{vol}(P)$. Also the second leading coefficient admits a simple geometric interpretation via the facets $F_1, \ldots, F_m$ of the lattice polytope. Here we have (cf. [24])

$$\mathrm{G}_{n-1}(P) = \frac{1}{2}\sum_{i=1}^{m} \frac{\mathrm{vol}_{n-1}(F_i)}{\det(\mathrm{aff}\,F_i \cap \mathbb{Z}^n)}, \tag{5.2}$$

where $\mathrm{vol}_{n-1}()$ denotes the $(n-1)$-dimensional volume and $\det(\mathrm{aff}\,F_i \cap \mathbb{Z}^n)$ denotes the determinant of the $(n-1)$-dimensional sublattice of $\mathbb{Z}^n$ contained in the affine hull of the facet $F_i$. So $\mathrm{G}_{n-1}(P)$ may be regarded as the normalised surface area of $P$ with respect to $\mathbb{Z}^n$ and, obviously we have $\mathrm{G}_{n-1}(P) \leq (1/2)\,\mathrm{F}(K)$.

All other coefficients $\mathrm{G}_i(P)$, $1 \leq i \leq n-2$, have no such direct geometric meaning, except for special classes of polytopes (cf., e.g., [9,10,20,32,38,37,42,45]). Since its discovery the Ehrhart polynomial and its coefficients play an essential role in discrete geometry, geometry of numbers and combinatorics (cf., e.g., [26,27,31,36,53]). For instance, in [12] Betke and Kneser showed that the coefficients form a basis of all additive and unimodular invariant functionals on the space $\mathcal{P}^n$. Stanley studied the Hilbert series of an integral polytope and proved in this context his famous non-negativity theorem [52]. For representations of $\mathrm{G}_i(P)$ in terms of Todd classes of a toric variety associated with $P$ we refer to [7] and the references within. Based on Barvinok's methods for counting lattice points (cf., e.g., [5,6]), De Loera et al. developed an efficient algorithm for calculating the coefficients of the Ehrhart polynomial (cf. [18,19]).

Before we present a relation between the coefficients of the Ehrhart polynomial and the successive minima we give some examples of Ehrhart polynomials. For the 3-dimensional Reeve simplex $R_3(m)$ (cf. (5.1)) we obtain by calculating the lattice points in $k\,R_3(m)$, $k = 1, 2, 3$, that

$$\mathrm{G}(k\,R_3(m)) = \frac{m}{6}\,k^3 + k^2 + \frac{12-m}{6}\,k + 1.$$

In particular, this shows that some of the $\mathrm{G}_i$'s might be negative. For the simplex $T_n = \mathrm{conv}\{0, e_1, \ldots, e_n\}$ and the cross-polytope $C_n^\star$ one easily finds

$$\mathrm{G}(k\,T_n) = \binom{n+k}{n}, \quad \mathrm{G}(k\,C_n^\star) = \sum_{i=0}^{n} 2^i \binom{n}{i}\binom{k}{i}.$$

If $Q$ is a lattice box defined via integers $\alpha_i$ (cf. (1.4)) we have

$$\mathrm{G}(k\,Q) = \prod_{i=1}^{n} (1 + 2\,\alpha_i\,k) \tag{5.3}$$

and, in particular, for the cube $C_n$ we obtain

$$\mathrm{G}_i(C_n) = 2^{n-i}\binom{n}{i}.$$

In recent years the Ehrhart polynomial was not only regarded as a polynomial for integers $k$, but as a formal polynomial of a complex variable $s \in \mathbb{C}$ (cf. [8,47,57]). Therefore, for $P \in \mathcal{P}^n$ and $s \in \mathbb{C}$ we set

$$\mathrm{G}(s, P) = \sum_{i=0}^{n} \mathrm{G}_i(P)\, s^i = \prod_{i=1}^{n}\left(1 + \frac{s}{\gamma_i(P)}\right),$$

where $-\gamma_i(P) \in \mathbb{C}$, $1 \leq i \leq n$, are the roots of the Ehrhart polynomial $\mathrm{G}(s, P)$. So $\mathrm{G}_i(P)$ is the $i$-th elementary symmetric polynomial of $1/\gamma_1, \ldots, 1/\gamma_n$ and, in particular, we have

$$\mathrm{vol}(P) = \mathrm{G}_n(P) = \prod_{i=1}^{n}\frac{1}{\gamma_i(P)} \quad \text{and} \quad \mathrm{G}_{n-1}(P) = \sum_{j=1}^{n}\prod_{i\neq j}\frac{1}{\gamma_i(P)}. \tag{5.4}$$

For bounds and further properties of the roots of the Ehrhart polynomial we refer to [8,30]. Here we are interested in relations between $\gamma_i(P)$ and Minkowski's successive minima. For a lattice box $Q$ as considered in (5.3) we obviously have

$$\frac{\lambda_i(Q)}{2} = \frac{1}{2\,\alpha_i} = \gamma_i(Q),\ 1 \leq i \leq n.$$

In terms of the roots of the Ehrhart polynomial (cf. (5.4)) we can rewrite Minkowski's second Theorem 1.2 for $P \in \mathcal{P}^n \cap \mathcal{K}_0^n$ as

$$\left(\prod_{i=1}^{n}\frac{\lambda_i(P)}{2}\right)^{1/n} \leq \left(\prod_{i=1}^{n}\gamma_i(P)\right)^{1/n} \leq n!^{1/n}\left(\prod_{i=1}^{n}\frac{\lambda_i(P)}{2}\right)^{1/n}. \tag{5.5}$$

These inequalities between the geometric mean of the successive minima and the roots of the Ehrhart polynomial lead naturally to the question of further inequalities among the elementary symmetric functions of $\lambda_i(P)/2$ and $\gamma_i(P)$. So far the only other known relation is between the arithmetic means [30]

**Theorem 5.3 (H., Schürmann, W.).** *Let $P \in \mathcal{P}^n \cap \mathcal{K}_0^n$. Then*

$$\frac{1}{n}\left(\sum_{i=1}^{n}\gamma_i(P)\right) \leq \frac{1}{n}\left(\sum_{i=1}^{n}\frac{\lambda_i(P)}{2}\right) \tag{5.6}$$

*and the bound is tight.*

An interesting feature of this inequality is the fact that it is tight for the cube $C_n$ as well as for the cross polytope $C_n^*$. However, there does not seem to be a lower bound on the arithmetic mean of $\gamma_i(P)$ by the arithmetic mean of $\lambda_i(P)$ (cf. [30]). In terms of the coefficients of the Ehrhart polynomial, Theorem 5.3 is equivalent to the more geometric statement (cf. (5.4))

$$\frac{\mathrm{G}_{n-1}(P)}{\mathrm{vol}(P)} \leq \frac{1}{2} \sum_{i=1}^{n} \lambda_i(P). \tag{5.7}$$

In comparison with (3.2) we see that the ratio of surface area to volume can be bounded from below by the successive minima whereas the ratio of the lattice surface area to the volume can be bounded form above. From (5.7) and the upper bound in (1.3) we get

**Corollary 5.4.** Let $P \in \mathcal{P}^n \cap \mathcal{K}_0^n$. Then

$$\mathrm{G}_{n-1}(P) \leq \sum_{j=1}^{n} \prod_{i \neq j} \frac{2}{\lambda_i(P)}.$$

This corollary brings us back to the polynomial

$$\mathrm{L}(\mu\, K) = \prod_{i=1}^{n} \left( \frac{2}{\lambda_i(K)} \mu + 1 \right)$$

introduced in (4.2). Denoting the coefficients of this polynomial by $\mathrm{L}_i(K)$ the right hand side in the corollary above is exactly $\mathrm{L}_{n-1}(P)$ and the leading coefficient is equal to $\prod_{i=1}^{n} 2/\lambda_i(P)$. Thus, by the upper bound in Minkowski's second theorem and Corollary 5.4 we obtain for $P \in \mathcal{P}^n \cap \mathcal{K}_0^n$ the two inequalities

$$\mathrm{G}_n(P) \leq \mathrm{L}_n(P) \text{ and } \mathrm{G}_{n-1}(P) \leq \mathrm{L}_{n-1}(P).$$

These two relations also support the conjectured inequality $G(K) \leq \mathrm{L}(K)$ for $K \in \mathcal{K}_0^n$ which follows from the stronger Conjecture 4.3. Observe, that it is enough to prove $\mathrm{G}(K) \leq \mathrm{L}(K)$ for 0-symmetric lattice polytopes. It seems to be quite unlikely, however, to prove $\mathrm{G}(P) \leq \mathrm{L}(P)$ via the associate polynomials in a coefficient-wise way. Nevertheless for special polytopes this approach may work. For instance, let $P$ be a 0-symmetric $n$-dimensional lattice polytope with $\mathrm{int}(P) \cap \mathbb{Z}^n = \{0\}$ and thus $\lambda_i(P) = 1$, $1 \leq i \leq n$. Then $\mathrm{L}_i(P) = \binom{n}{i} 2^i$ and $\mathrm{G}_i(P) \leq \mathrm{L}_i(P)$ is equivalent to a conjecture of Wills [56]

*Conjecture* 5.5 (W.). Let $P \in \mathcal{P}^n \cap \mathcal{K}_0^n$ such that $\mathrm{int}(P) \cap \mathbb{Z}^n = \{0\}$. Then

$$\mathrm{G}_i(P) \leq \binom{n}{i} 2^i, \quad 1 \leq i \leq n.$$

Of course, the case $i = n$ follows from Minkowski's first Theorem 1.1.

## 6. Best simultaneous Diophantine approximations

Here we want to extend Minkowski's theorems on successive minima to periodic lattices. As in the previous sections we will present the results only with respect to the integral lattice, but as before they can easily be generalised to aribitrary lattices. For $\alpha \in \mathbb{R}^n$ and an integer $Q \in \mathbb{N}_{\geq 0}$ we set

$$\mathbb{Z}^n(\alpha, Q) = \mathbb{Z}^n \cup (\alpha + \mathbb{Z}^n) + (2\alpha + \mathbb{Z}^n) \cup \cdots \cup (Q\alpha + \mathbb{Z}^n),$$

where we always assume that $k\alpha \notin \Lambda$ for $1 \leq k \leq Q$. In order to allow the case $\alpha \in \Lambda$ we admit $Q = 0$. $\mathbb{Z}^n(\alpha, Q)$ is called a periodic lattice. Next we define for $K \in \mathcal{K}_0^n$ and $1 \leq i \leq n$

$$\lambda_i(K, \mathbb{Z}^n(\alpha, Q)) = \min\{\lambda > 0 : \dim(\lambda K \cap \mathbb{Z}^n(\alpha, Q)) \geq i\}.$$

Obviously, for $Q = 0$ we just have Minkowski's successive minima introduced in Section 1. Denoting by $|x|_K = \min\{\lambda \geq 0 : x \in \lambda K\}$ the norm of $x \in \mathbb{R}^n$ induced by $K$, the first successive minimum of the series above can be expressed by

$$\lambda_1(K, \mathbb{Z}^n(\alpha, Q)) = \min\{|q\alpha - z|_K > 0 : q \in \{0, \ldots, Q\}, z \in \mathbb{Z}^n\}.$$

So it measures the quality of a best approximation of $\alpha$ by a rational vector whose common denominator is bounded by $Q$. This functional has been studied from various respects. For instance, for $n = 1$ and based on continued fractions, Klein [34] gave a geometric interpretation of such a "best approximation point" as a vertex of an associated 2-dimensional Klein polyhedron. Davenport and Mahler [17] proved that there exist infinitely many points $(q, z)^{\intercal} \in \mathbb{Z}^3$, $z \in \mathbb{Z}^2$, such that $|q\alpha - z|_{B^2}^2 \leq (2/\sqrt{23})/q$ and the constant on the right hand side is best possible. The first who embedded $\lambda_1(K, \mathbb{Z}^n(\alpha, Q))$ in a series of functionals were W. B. Jurkat [33] and W. Kratz [35]. Their functionals are closely linked to the $\lambda_i(K, \mathbb{Z}^n(\alpha, Q))$ but defined in a space of dimension $n + 1$.

In order to present Minkowski-type inequalities for the $\lambda_i(K, \mathbb{Z}^n(\alpha, Q))$ we also need the notation of the density of a densest (not necessarily lattice) packing of $K$ which will be denoted by $\delta^*(K)$ (cf [27, pp. 223]). Then we have the following analogues to Minkowski's theorems [2]

**Theorem 6.1 (Aliev, H.).** *Let $K \in \mathcal{K}_0^n$, $\alpha \in \mathbb{R}^n$ and $Q \in \mathbb{N}_{\geq 0}$ such that $k\alpha \notin \Lambda$ for $1 \leq k \leq Q$. Then with $\lambda_i = \lambda_i(K, \mathbb{Z}^n(\alpha, Q))$ we have*

$$\text{i)} \quad (\lambda_1)^n \mathrm{vol}(K) \leq \delta^*(K)\, 2^n \frac{1}{Q+1},$$

$$\text{ii)} \quad \gamma(\alpha, Q, n) \frac{2^n}{n!} \leq \lambda_1 \cdot \ldots \cdot \lambda_n \,\mathrm{vol}(K) \leq 2^n \frac{1}{Q+1},$$

*where $\gamma(\alpha, Q, n)$ is a certain constant depending on $\alpha$, $Q$ and $n$.*

Inequalities of that type give us information on the quality of the simultaneous approximation of a vector by a system of linearly independent rational vectors whose common denominators are bounded.

For details on the constant $\gamma(\alpha, Q, n)$ in the lower bound of Theorem 6.1 ii) we refer to the paper [2], but we want to remark that in the case $(Q+1)\alpha \in \mathbb{Z}^n$, i.e., $\mathbb{Z}^n(\alpha, Q)$ is a lattice, we have $\gamma(\alpha, Q, n) \geq 1/(Q+1)$. Thus Theorem 6.1 may be regarded as an extension of Minkowski's inequalities (1.2) and (1.3) to periodic lattices. Actually, the first inequality i) looks quite similar to (2.1), but since the inequality is valid for a more general structure than lattices we have to replace the density of densest lattice packing in (2.1) by $\delta^\star(K)$.

If $K = C_n$ then $|\cdot|_K$ is the maximum norm and Theorem 6.1 i) says that for any $\alpha \in \mathbb{R}^n$ there exists a $z \in \mathbb{Z}^n$ and $q \in \{1, \ldots, Q\}$ such that

$$\left|\alpha_i - \frac{z_i}{q}\right| < \frac{Q^{-1/n}}{q}, \quad 1 \leq i \leq n.$$

This is Dirichlet's classical theorem on simultaneous Diophantine approximation.

Quite recently Aliev and Gruber [1] also gave a lower estimate for the quantity $\lambda_1(K, \mathbb{Z}^n(\alpha, Q))$ which, with the notation of Theorem 6.1, can be formulated as

**Theorem 6.2 (Aliev, Gruber).** *For every $\epsilon > 0$ there exists an $\alpha \in \mathbb{R}^n$ and $Q \in \mathbb{N}$ such that*

$$(\lambda_1)^n \operatorname{vol}(K) > \delta(K)\, 2^n \frac{1}{Q}.$$

In fact, they proved this theorem not only for 0-symmetric convex bodies, but for any bounded star body $K$.

## References


[1] I. Aliev and P. M. Gruber, Best simultaneous Diophantine approximations under a constraint on the denominator, *Contrib. Discrete Math.* **1** (2006) no. 1, 29–46

[2] I. Aliev and M. Henk, Successive minima and best simultaneous diophantine approximations, *Mh. Math.* **147** (2006) no. 2, 95–101

[3] R. P. Bambah, A. C. Woods, and H. Zassenhaus, Three proofs of Minkowski's second inequality in the geometry of numbers, *J. Austral. Math. Soc.* **5** (1965) 453–462

[4] R. P. Bambah, V. C. Dumir, and R. J. Hans-Gill, Diophantine inequalities, *Proc. Natl. Acad. Sci. India, Sect. A* **68** (1998) no. 2, 101–114, *corrigendum ibid.* **69**, no. 1, 103 (1999)

[5] A. Barvinok, Computing the Ehrhart polynomial of a convex lattice polytope, *Discrete Comput. Geom.* **12** (1994) no. 1, 35–48

[6] A. Barvinok, A polynomial time algorithm for counting integral points in polyhedra when the dimension is fixed, *Math. Oper. Res.* **19** (1994) 769–779

[7] A. Barvinok and J. E. Pommersheim, An algorithmic theory of lattice points in polyhedra, New Perspectives in Algebraic Combinatorics (Berkeley, CA, 1996–1997), *Math. Sci. Res. Inst. Publ.*, vol. 38, 1994, pp. 91–147

[8] M. Beck, J. De Loera, M. Develin, J. Pfeifle and R. P. Stanley, Coefficients and roots of Ehrhart polynomials, *Contemp. Math.* **374** (2005) 15–36

[9] M. Beck and D. Pixton, The Ehrhart polynomial of the Birkhoff polytope, *Discrete Comput. Geom.* **30** (2003) no. 4, 623–637

[10] U. Betke and P. Gritzmann, An application of valuation theory to two problems of discrete geometry, *Discrete Math.* **58** (1986) 81–85

[11] U. Betke, M. Henk, and J. M. Wills, Successive-minima-type inequalities, *Discrete Comput. Geom.* **9** (1993) no. 2, 165–175

[12] U. Betke and M. Kneser, Zerlegungen und Bewertungen von Gitterpolytopen, *J. Reine Angew. Math.* **358** (1985) 202–208

[13] J. Bochnak and W. Kucharz, On successive minima of indefinite quadratic forms, *Enseign. Math. (2)* **51** (2005) no. 3–4, 319–330

[14] J. G. van der Corput, Verallgemeinerung einer Mordellschen Beweismethode in der Geometrie der Zahlen, *Acta Airithm.* **1** (1935) 62–66

[15] H. Davenport, Minkowski's inequality for the minima associated with a convex body, *Quarterly J. Math.* **10** (1939) 119–121

[16] H. Davenport, The product of $n$ homogeneous linear forms, *Indag. Math.* **8** (1946) 525–531

[17] H. Davenport and K. Mahler, Simultaneous diophantine approximation, *Duke Math. J.* **13** (1946) 105–111

[18] J. De Loera, D. Haws, R. Hemmecke and P. Huggins, A user's guide for `latte` v1.1, software package `latte`, 2004, available at `http://www.math.ucdavis.edu/~latte`

[19] J. De Loera, R. Hemmecke, J. Tauzer and R. Yoshida, *Effective lattice point counting in rational convex polytopes*, *J. Symb. Comput.* **38** (2004) no. 4, 1273–1302

[20] R. Diaz and S. Robins, The Ehrhart polynomial of a lattice polytope, *Ann. of Math.* **145** (1997) no. 3, 503–518, Erratum in 146:1 (1997) 237

[21] V. C. Dumir and R. J. Hans-Gill, The second minimum for positive values of non-homogeneous ternary quadratic forms of type (1, 2)., *Ranchi Univ. Math. J.* **28** (1997) 65–75

[22] E. Ehrhart, Sur les ovales et les ovoides, *C. R. Acad. Sci. Paris* **258** (1955) 573–575

[23] E. Ehrhart, Sur les polyèdres rationnels homothétiques à n dimensions, *C. R. Acad. Sci., Paris, Sér. A* **254** (1962) 616–618

[24] E. Ehrhart, Sur un problème de géométrie diophantienne linéaire, *J. Reine Angew. Math.* **227** (1967) 25–49

[25] E. Ehrhart, Sur la loi de réciprocité des polyèdres rationnels, *C. R. Acad. Sci., Paris, Sér. A* **266** (1968) 695–697

[26] P. Gritzmann and J. M. Wills, Lattice points, Handbook of convex geometry (P.M. Gruber and J.M. Wills, eds.), vol. B, North-Holland, Amsterdam, 1993

[27] P. M. Gruber and C. G. Lekkerkerker, Geometry of numbers, second ed., vol. 37, North-Holland Publishing Co., Amsterdam, 1987

[28] M. Henk, Inequalities between successive minima and intrinsic volumes of a convex body, *Monatsh. Math.* **110** (1990) no. 3–4, 279–282

[29] M. Henk, Successive minima and lattice points, *Rend. Circ. Mat. Palermo (2) Suppl.* (2002) no. 70, part I, 377–384

[30] M. Henk, A. Schürmann, and J. M. Wills, Ehrhart polynomials and successive minima, Mathematika, (to appear)

[31] T. Hibi, Algebraic combinatorics on convex polytopes, Carslaw Publications, Glebe, Australia, 1992

[32] T. Hibi, Dual polytopes of rational convex polytopes, *Combinatorica* **12** (1992) no. 2, 237–240

[33] W. B. Jurkat, On successive minima with constraints, *Analysis* **1** (1981) 33–44

[34] F. Klein, Über eine geometrische Auffassung der gewöhnlichen Kettenbruchentwicklung, *Nachr. Ges. Wiss. Göttingen, Math.-Phys.* **3** (1895) 357–359

[35] W. Kratz, Sukzessive Minima mit und ohne Nebenbedingungen, *Mh. Math.* **91** (1981) 275–289

[36] J. C. Lagarias, Point lattices, Handbook of Combinatorics (R. L. Graham, M. Grötschel and L. Lovász, eds.), vol. A, North-Holland, Amsterdam, 1995

[37] F. Liu, Ehrhart polynomials of lattice-face polytopes, *Trans. Amer. Math. Soc.* **360** (2008) 3041–3069

[38] F. Liu, Ehrhart polynomials of cyclic polytopes, *J. Comb. Theory, Ser. A* **111** (2005) 111–127

[39] P. McMullen, Non-linear angle-sum relations for polyhedral cones and polytopes, *Math. Proc. Cambridge Philos. Soc.* **78** (1975) no. 2, 247–261

[40] H. Minkowski, Geometrie der Zahlen, Teubner, Leipzig-Berlin, 1896, Reprinted: Johnson, New York, 1968

[41] H. Minkowski, *Dichteste gitterförmige Lagerung kongruenter Körper*, Nachr. K. Ges. Wiss. Göttingen, Math.-Phys. KL (1904), 311–355, see Gesammelte Abhandlungen vol. II, 3–42, Leipzig 1911

[42] L. J. Mordell, Lattice points in tetrahedron and generalized Dedekind sums, *J. Indian Math. Soc. (N.S.)* **15** (1951) 41–46

[43] M. Ram Murty and N. Thain, Pick's theorem via Minkowski's theorem, *Amer. Math. Monthly* **114** (2007) no. 8, 732–736

[44] G. A. Pick, Geometrisches zur Zahlenlehre, *Sitzungsber. Lotus Prag* **19** (1899) 311–319

[45] J. E. Pommersheim, Toric varieties, lattice points and Dedekind sums, *Math. Ann.* **295** (1993) no. 1, 1–24

[46] J. E. Reeve, On the volume of lattice polyhedra, *Proc. London Math. Soc.* **7** (1957) no. 3, 378–395

[47] F. Rodriguez-Villegas, On the zeros of certain polynomials, *Proc. Amer. Math. Soc.* **130** (2002) 2251–2254

[48] R. Schneider, Convex bodies: The Brunn-Minkowski theory, Encyclopedia of Mathematics and its Applications, vol. 44, Cambridge University Press, Cambridge, 1993

[49] C. L. Siegel, Über gitterpunkte in konvexen körpern und ein damit zusammenhängendes Extremalproblem, *Acta Math.* **65** (1935) 307–323

[50] C. L. Siegel, Lectures on the geometry of numbers, Springer-Verlag, Berlin, 1989

[51] M. Sombra, Successive minima of projective toric varieties (Minimums successifs des variétés toriques projectives), *J. Reine Angew. Math.* **586** (2005) 207–233, (French)

[52] R. P. Stanley, Decompositions of rational convex polytopes, *Ann. Discrete Math.* **6** (1980) 333–342.

[53] R. P. Stanley, Enumerative combinatorics. vol. 1, Cambridge Studies in Advanced Mathematics, vol. 49, Cambridge University Press, Cambridge, 1997, Corrected reprint of the 1986 original

[54] J. L. Thunder, Remarks on adelic geometry of numbers., Number theory for the millennium III. Proceedings of the millennial conference on number theory, Urbana-Champaign, IL, USA, May 21–26, 2000 (M.A. Bennett et al, ed.), 2002, pp. 253–259

[55] H. Weyl, On geometry of numbers, *Proc. London Math. Soc. (2)* **47** (1942) 268–289

[56] J. M. Wills, On an analog to Minkowski's lattice point theorem, The Geometric Vein: The Coxeter Festschrift (C. Davis, B. Grünbaum and F. A. Sherk, eds.), Springer, New York, 1981, pp. 285–288

[57] J. M. Wills, Minkowski's successive minima and the zeros of a convexity-function, *Monatsh. Math.* **109** (1990) no. 2, 157–164

[58] A. C. Woods, The anomaly of convex bodies, *Proc. Cambridge Philos. Soc.* **52** (1956) 406–423

Proc. Int. Conf. – Number Theory and Discrete Geometry No. 6, 2008, pp. 143–147.

# A Formula Relating the Discriminant of Finite Extensions of Local Fields and Tignol's Constant

Sudesh K. Khanduja[1]* and Kaori Ota[2]†

[1]Department of Mathematics, Panjab University, Chandigarh 160 014, India
e-mail:skhand@pu.ac.in
[2]Department of Mathematics and Computer Science, Tsuda College, 2-1-1 Tsuda-cho, Kodaira-shi, Tokyo, 187-8577 Japan
e-mail: ota@tsuda.ac.jp

**Abstract.** Let $K$ be a local field with valuation $v$ having value group $\mathbb{Z}$. For a finite separable extension $(L, w)$ of $(K, v)$, let $A_{L/K}$ stand for the set $\{v(Tr_{L/K}(\alpha)) - w(\alpha) \mid 0 \neq \alpha \in L\}$. In 1990, Tignol gave a formula to calculate $\min A_{L/K}$ (referred to as Tignol's constant) when $L/K$ is an extension of degree equal to the characteristic of the residue field of $v$. In the present paper, for any finite separable extension $L/K$, it has been shown that

$$\min A_{L/K} = \frac{d(L/K)}{[L:K]} - 1 + \frac{1}{e(L/K)},$$

where the discriminant of $L/K$ is the ideal $P_K^{d(L/K)}$, $P_K$ being the maximal ideal of the valuation ring of $v$ and $e(L/K)$ the ramification index of $w/v$.

**Keywords.** Valued fields, local fields.

**2000 Subject Classification:** Primary 12J10, Secondary 11S15

*Dedicated to Professor R. P. Bambah on his 80th birthday*

## 1. Introduction

Let $v$ be a Henselian valuation of arbitrary rank of a field $K$ with value group $\Gamma_K$. A finite extension $(L, w)/(K, v)$ (or briefly $L/K$) is called defectless if $[L : K] = ef$ where $e$ and $f$ are the index of ramification and the residual degree of $w/v$. In 1990, Tignol proved that if $(L, w)/(K, v)$ is a finite separable defectless extension of degree $p$ equal to the characteristic of the residue field of $v$, then the set $A_{L/K}$ defined by

$$A_{L/K} = \{v(Tr_{L/K}(\alpha)) - w(\alpha) \mid \alpha \in L, \ \ \alpha \neq 0\} \tag{1}$$

has a minimum element (cf. [4, Prop. 2.5 ]). Indeed he has calculated the constant $\min A_{L/K}$ (referred to as Tignol's constant) by showing that if $\theta$ is an element of $L$

*All correspondence may be addressed to this author.

†After this paper was written, we found that Hyodo had already considered $\min A_{L/K}$ and named it the depth of ramification for an $n$ dimensional complete discrete valuation field $K$ in [5]. In the paper he mentioned the above formula for $\min A_{L/K}$, but no proof was given.

having the minimum polynomial $f(x)$ over $K$ and satisfying either $w(\theta)$ is not in $\Gamma_K$ or $w(\theta) = v(a)$ is in $\Gamma_K$, $a \in K$, with the $w$-residue of $\theta/a$ not belonging to the residue field of $v$, then

$$\min A_{L/K} = w(f'(\theta)) - (p-1)w(\theta). \tag{2}$$

In 2004, Khanduja and Singh [2] characterized defectless extensions by proving that a finite separable extension $(L, w)$ of a Henselian valued field $(K, v)$ is defectless if and only if the set $A_{L/K}$ defined by (1) has a minimum element. They also proved that for a tower $K \subset M \subset L$ of finite separable defectless extensions of Henselian valued fields, Tignol's constant satisfies

$$\min A_{L/K} = \min A_{L/M} + \min A_{M/K}.$$

The following question naturally arises. Can we find a simple formula like the one given by (2) for $\min A_{L/K}$, where $L$ is any finite separable defectless extension of a Henselian valued field $(K, v)$? In this paper, our aim is to give such a formula when $K$ is a local field, i.e., $K$ is a finite extension of the field $\mathbf{Q}_p$ of $p$-adic numbers or of $F_p((t))$.

In what follows, for a finite extension $(L, w)/(K, v)$ of local fields, $d(L/K)$ will denote the non-negative integer defined by $P_K^{d(L/K)} = disc(L/K)$, where $P_K$ is the maximal ideal of the valuation ring $O_K$ of $v$ and $disc(L/K)$ stands for the discriminant of $L/K$ defined in the usual way. Also $e(L/K)$ and $f(L/K)$ will stand respectively for the index of ramification and the residual degree of $(L, w)/(K, v)$. With the above notations, we prove

**Theorem 1.1.** *Let $K$ be a local field with valuation $v$ having value group $\mathbb{Z}$. Let $(L, w)$ be a finite separable extension of $(K, v)$. Then*

$$\min A_{L/K} = \frac{d(L/K)}{[L:K]} - 1 + \frac{1}{e(L/K)}.$$

It may be pointed out that in case $L/K$ is an extension of degree equal to the characteristic of the residue field of $v$, then the formula given in the above theorem coincides with the one given by Tignol in (2) in view of Proposition 2.A(ii) stated in the following section.

## 2. Some preliminary results

The following proposition is well known (cf. [3, Chapter III]).

**Proposition 2.A.** *Let $(K, v)$ be as in Theorem 1.1 and $(L, w)/(K, v)$ be a finite separable extension. Let $M/K$ be a subextension of $L/K$. Then the following hold:*

(i) $d(L/K) = [L:M]d(M/K) + f(M/K)d(L/M)$.
(ii) *If $O_L = O_K[\theta]$ and $f(X)$ is the minimal polynomial of $\theta$ over $K$, then*

$$d(L/K) = [L:K]\,w(f'(\theta)).$$

Recall that if $(L, w)/(K, v)$ be an extension of degree $n$, a basis $\theta_1, \ldots, \theta_n$ of $L/K$ is called a valuation basis (with respect to $w/v$) if for each $\theta = \sum a_i\theta_i$ in $L$, $a_i \in K$, the equation $w(\theta) = \min_i w(a_i\theta_i)$ holds.

The lemma stated below is already known. Its proof is omitted (cf. [4, p.11], [2, Lemma 2.2])

**Lemma 2.B.** *If* $\{\alpha_1, \ldots, \alpha_n\}$ *is a valuation basis of a separable extension* $(L, w)/(K, v)$ *of degree* $n$, *then*

$$\min A_{L/K} = \min\{v(Tr_{L/K}(\alpha_i)) - w(\alpha_i) \mid 1 \leq i \leq n\}.$$

We also need the following already known results (cf. [1], [2, Theorem 2.5]).

**Theorem 2.C.** *Let* $(L, w)$ *be a finite separable extension of a Henselian valued field* $(K, v)$.

(i) *If* $L/K$ *is an unramified extension, then* $\min A_{L/K}$ *is zero.*
(ii) *For any subextension* $M/K$ *of* $L/K$, *one has*

$$\min A_{L/K} = \min A_{L/M} + \min A_{M/K}.$$

## 3. Proof of Theorem 1.1

Let $p$ be the characteristic of the residue field of $v$. The theorem is proved in two steps.

**Step I.** We first prove the theorem when $L/K$ is a separable totally ramified extension of degree $n$.

Let $\pi$ be a prime element of $L$. In the present case $w(\pi) = \frac{1}{n}$ and $O_L = O_K[\pi]$, so by Proposition 2.A(ii), we obtain

$$d(L/K) = [L : K]\, w(f'(\pi)), \tag{3}$$

where $f(X)$ is the minimal polynomial of $\pi$ over $K$. Since $\{1, \pi, \ldots, \pi^{n-1}\}$ is a valuation basis for $L/K$, so is $\left\{\frac{1}{f'(\pi)}, \frac{\pi}{f'(\pi)}, \ldots, \frac{\pi^{n-1}}{f'(\pi)}\right\}$. Hence from Lemma 2.B, we have

$$\min A_{L/K} = \min\left\{v\left(\mathrm{Tr}_{L/K}\left(\frac{\pi^i}{f'(\pi)}\right)\right) - w\left(\frac{\pi^i}{f'(\pi)}\right) |0 \leq i \leq n-1\right\}.$$

But it is well known that (cf. [3, Chp. 3, Lemma 2])

$$\mathrm{Tr}_{L/K}\left(\frac{\pi^i}{f'(\pi)}\right) = 0 \quad \text{for} \quad 0 \leq i \leq n-2 \quad \text{and} \quad \mathrm{Tr}_{L/K}\left(\frac{\pi^{n-1}}{f'(\pi)}\right) = 1.$$

Therefore

$$\min A_{L/K} = -w\left(\frac{\pi^{n-1}}{f'(\pi)}\right) = w(f'(\pi)) - 1 + \frac{1}{n};$$

the above equation together with (3) proves the required formula in the present case.

**Step II.** We now derive the theorem when $L/K$ is any finite separable extension. Take the maximal unramified subextension $L_0/K$ of $L/K$. Keeping in mind $d(L_0/K) = 0$, it follows from Proposition 2.A(i) that

$$d(L/K) = f(L/K)\,d(L/L_0).$$

Also from Theorem 2.C and Step I applied to $L/L_0$, we have

$$\begin{aligned}\min A_{L/K} &= \min A_{L/L_0} + \min A_{L_0/K} = \min A_{L/L_0}\\ &= \frac{1}{[L:L_0]}d(L/L_0) - 1 + \frac{1}{[L:L_0]}\\ &= \frac{1}{[L:K]}d(L/K) - 1 + \frac{1}{e(L/K)}\end{aligned}$$

as desired.

*Remark.* Let $(K, v)$ be as above and let $L/K$ be a finite separable totally ramified extension of degree $n$, $\pi$ be a prime element of $L$ and

$$f(X) = X^n + a_1X^{n-1} + \cdots + a_iX^{n-i} + \cdots + a_n$$

the minimal polynomial of $\pi$ over $K$. Take the integer $j$ satisfying

$$w(f'(\pi)) = w((n-j)a_j\pi^{n-j-1}) \qquad \text{with } 0 \leq j \leq n-1.$$

Using Proposition 2.A(ii) and Theorem 1.1, one can quickly verify that

$$\min A_{L/K} = v(n-j) + v(a_j) - \frac{j}{n}.$$

But by Lemma 2.B, one has

$$\min A_{L/K} = \min_{0\leq i\leq n-1}\{v(Tr_{L/K}(\pi^i)) - w(\pi^i)\}.$$

The above two equations yield the following interesting equalities

$$\min A_{L/K} = v(Tr_{L/K}(\pi^j)) - w(\pi^j)$$

and

$$v(Tr_{L/K}(\pi^j)) = v(n-j) + v(a_j).$$

## Acknowledgements

This paper was written while the second author was visiting Department of Mathematics at Panjab University, Chandigarh. She is highly grateful to the first author and the department for the support and the hospitality given to her. The research of the first author is partially supported by CSIR, India.

## References

[1] S. K. Khanduja, A Characterization of Finite Tame Extensions, Bulletin of London Math. Society **32** (2000) 551–554

[2] S. K. Khanduja, A. P. Singh, On a Theorem of Tignol for Defectless extensions and its converse, *J. Algebra* **288** (2005) 400–408

[3] J. -P. Serre, Local Fields, Graduate Texts in Mathematics **67** Springer-Verlag (1979)

[4] J. -P. Tignol, Algèbras à division et extensions de corps sauvagement ramifiées de degré premier, *J. reine angew. Math.* **404** (1990) 1–38

[5] O. Hyodo, Wild ramification in the imperfect residue field case, *Advanced Studies in Pure Mathematics* **12** (1987), *Galois Representations and Arithmetic Algebraic Geometry*, 287–314

Proc. Int. Conf. – Number Theory and Discrete Geometry No. 6, 2008, pp. 149–154.

# Vacancy Phenomena in Finite Sphere Packings

Wlodzimierz Kuperberg

Department of Mathematics and Statistics, Auburn University, Auburn, AL 36849-5310
e-mail: kuperwl@auburn.edu

**Abstract.** What is the maximum radius $r_k$ of $k$ congruent balls that can be packed in a unit spherical container in $\mathbb{R}^n$ ($n \geq 2$)? Obviously, $r_{k+1} \leq r_k$ for every $k$. H. Davenport and G. Hajós [DH], and, independently, R. A. Rankin [R], proved that $r_{2n} = r_{n+2}$ for every $n \geq 2$, that is, if the container is large enough to hold $n+2$ $n$-dimensional balls of radius $r$ each, then there is room in it for as many as $2n$ balls of the same size—a vacancy phenomenon. We are concerned with the occurrences of such vacancy phenomena and with their extent, and we prove that if the equality $r_{k+s} = r_k$ occurs, then $s \leq k-2$. We also list a few examples of similar phenomena that occur in packing balls into some other convex containers.

## 1. Introduction and preliminaries

Throughout this note, an ($n$-dimensional) *convex body* is a compact convex subset of $\mathbb{R}^n$ with a nonempty interior. A collection of convex bodies is a *packing* if their interiors are mutually disjoint.

A typical finite packing problem asks for the maximum number of congruent copies of a given convex body $K$ that can be packed in a given, usually convex as well, container $C$. More specifically, one asks for the maximum size of $k$ congruent replicas of a body $K'$ homothetic to $K$, that can be packed in the given container. The main focus of this note is on the problem of finite packings of congruent $n$-dimensional balls. Let $r_k(C)$ denote the maximum radius of $k$ congruent balls that can be packed in the container $C$. (Obviously, the configuration of the balls' centers in the container determines the packing of the balls.)

The *inradius* of a convex body $C$, denoted by $r(C)$, is the maximum radius of a ball contained in $C$. For any positive number $a < r(C)$, the set of centers of balls of radius $a$ that are contained in $C$ is a convex body, called *$a$-inner body* of $C$, and denoted by $C_{-a}$.

For any finite set $F \subset \mathbb{R}^n$ of cardinality $k \geq 2$, the minimum distance between points in $F$ is called the *dispersion* of $F$ and denoted by $\delta(F)$ (not to be confused with other notions under the same name, found in statistics and in some other areas of mathematics). Given a convex body $C$ and an integer $k \geq 2$, among all $k$-point subsets of $C$ there is one of maximum dispersion. This maximum dispersion is denoted by $\delta_k(C)$, and each subset $F$ whose dispersion attains the maximum is said to be an *optimal set*, or an *optimal configuration* of $k$ points in $C$.

If $k \geq 2$ balls of radius $r = r_k(C)$ are packed in a container $C$, then, by definition of the inner body, the centers of the balls lie in $C_{-r}$, and the distance between any two centers is at least $2r$. Conversely, if $k$ points can be selected in $C_{-r}$ so that the distance between any two of them is at least $2r$, then these points are the centers of $k$ balls of radius $r$ each, packed in $C$. It follows that $r_k(C) = \frac{1}{2}\delta_k(C_{-r})$. Thus, of special interest to us are containers $C$ that possess the following property:

**Property (*).** For every $r < r(C)$ the inner body $C_{-r}$ is homothetic to $C$.

For containers $C$ with property $(*)$, the problem of finding the maximum radius of $k$ congruent balls that can be packed in the container $C$, and describing all possible configurations of the balls, is equivalent to the problem of finding the number $\delta_k(C)$, and describing all optimal configurations of $k$ points in $C$, respectively. The relation between $r_k(C)$ and $\delta_k(C)$ is:

$$r_k(C) = \frac{r(C)\delta_k(C)}{2r(C) + \delta_k(C)}.$$

Observe that if $C$ is a polyhedron, then $C$ satisfies $(*)$ if and only if $C$ is circumscribed about a ball. Hence property $(*)$ can be considered as a generalization of the property of being circumscribed about a ball. This comment is justified further by the following, general construction:

Choose a subset $A$ (finite or infinite) of a sphere $S$ in $\mathbb{R}^n$ such that the interior of the convex hull of $A$ contains the sphere's center. For each point $x \in A$ let $h_x$ denote the closed half-space determined by the plane tangent to $S$ at $x$ and containing $S$. The intersection $C$ of all of these half-spaces, $C = \bigcap_{x \in A} h_x$, is a convex body, and it is easy to show that $C$ satisfies condition $(*)$ (the inradius $r(C)$ is the radius of $S$). Moreover, it is not difficult to prove that the above construction produces *all* convex bodies $C$ that satisfy condition $(*)$. Obviously, a suitably chosen *finite* set $A$ yields any given convex polyhedron circumscribed about a ball.

For a similar, but more detailed analysis of the types of containers for which the problems of packing and of dispersion are equivalent, see Melissen [M, Chapter 3]. Although Melissen's analysis is restricted to the plane, many of his observations can be immediately generalized to all dimensions.

We assume the following notation:

$B^n$–the unit ball in $\mathbb{R}^n$;
$I^n$–the $n$-dimensional cube $[-1, 1]^n$;
$X^n$–the polytope dual to the $n$-cube, also called the regular $n$-dimensional *crosspolytope*.

Each of the containers considered in the following sections: the ball, the cube, and the regular crosspolytope, obviously satisfies condition $(*)$, hence in each case we switch from the packing problem to the dispersion problem immediately.

## 2. Packing balls in a spherical container

Here we assume that the container is spherical, say $C$ is the unit ball $B^n$ in $\mathbb{R}^n$, and we pack congruent balls in it. The corresponding dispersion problem is: Find the maximum

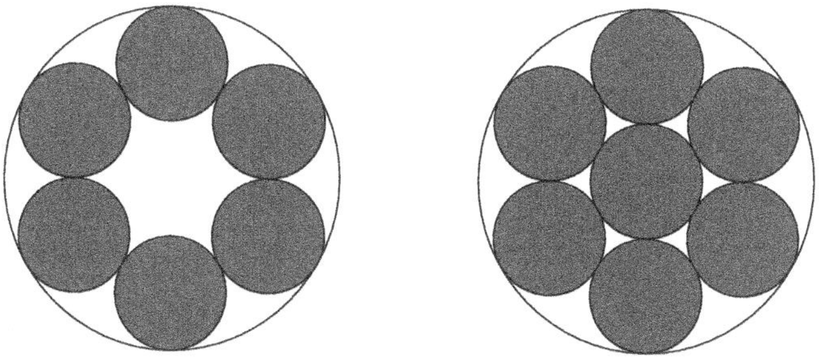

**Figure 1.** In any optimal packing of six circles, the container can hold seven.

number $\delta = \delta_k(B^n)$ such that $k$ points can be placed in $B^n$ so that each distance between the points is at least $\delta$.

The problem in its full generality seems prohibitively difficult; the exact values of $\delta_k(B^n)$ are known in a relatively few cases only. The values of $\delta_k(B^2)$ (in the plane) are known for $k \leq 14$ and $k = 19$ only (see [M] and [F99,F00,F03,F04]). In dimensions $n \geq 3$, the values of $\delta_k(B^n)$ are known for all $k \leq 2n$ (see [R] and [DH]) plus a few (finitely many) additional cases that depend on our knowledge of the Newton number, also known as the "kissing number," of the ball (see [CS]).

Obviously, $\delta_{k+1}(B^n) \leq \delta_k(B^n)$, and an occasional occurrence of equality is not surprising. It is easy to see, for example, that $\delta_6(B^2) = \delta_7(B^2) = 1$ (see Fig. 1). Each occurrence of equality $\delta_{k+1}(B^n) = \delta_k(B^n)$ shall be called a *vacancy phenomenon*, for it means that if the container is large enough to hold $k$ congruent balls, then there is enough room in it for at least one more ball of the same size.

In two dimensions (packing circles in a circle) the only proven instance of vacancy is the above-mentioned case $\delta_6(B^2) = \delta_7(B^2)$. The equality $\delta_{18}(B^2) = \delta_{19}(B^2)$, while highly believable (see Fig. 2), is still but a conjecture.

It is somewhat unexpected that, at least in high dimensions, a vacancy can be quite extensive, with "extra room" large enough to accommodate a good number of additional balls of the same size. If the equality $\delta_{k+s}(B^n) = \delta_k(B^n)$ occurs, let us say that we have a *vacancy of size s*, and that its *relative size* is $s/k$. In every dimension $n \geq 3$, a

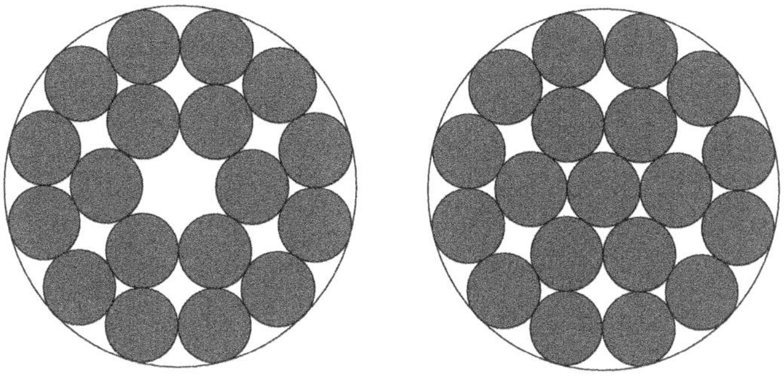

**Figure 2.** The conjectured vacancy $r_{18}(B^2) = r_{19}(B^2)$.

striking vacancy phenomenon occurs. According to a theorem of Davenport and Hajós [DH] (see also [R] and [AS]),

$$\delta_{n+2}(B^n) = \delta_{2n}(B^n) = \sqrt{2} \quad \text{for every } n \geq 2.$$

In other words, in $n$ dimensional space ($n \geq 3$) a vacancy of size $n-2$ and of relative size $\frac{n-2}{n+2}$ occurs, which shows that the vacancy size can be arbitrarily large, and its relative size can be arbitrarily close to 1 in sufficiently high dimensions. Moreover, Rankin's theorem specifies the optimal $2n$-point subset of the unit ball in $\mathbb{R}^n$: the points form the set of vertices of a regular $n$-dimensional regular crosspolytope (the $n$-crosspolytope, for short) inscribed in the ball. For $n+2 \leq k < 2n$, the optimal configurations are not unique.

The main goal of this note is to prove that the relative size of a vacancy for packing balls in a spherical container cannot ever reach or exceed 1. More precisely, we assert the following:

**Theorem 1.** *If* $\delta_{k+s}(B^n) = \delta_k(B^n)$, *then* $s \leq k-2$.

*Proof.* We will prove that $\delta_{2k-1}(B^n) < \delta_k(B^n)$, which is equivalent to the statement of the theorem. Let $F$ be an optimal set of $2k-1$ points in $B^n$, let $S$ denote the boundary sphere of $B^n$, and let $T$ be the set of points in $F$ that lie on $S$. Let $H$ be an $(n-1)$-dimensional hyperplane through the center $O$ of $S$ that contains no point from $T$. Then one of the half-spaces defined by $H$ contains at most $k-1$ points from $T$. Deleting these points from $F$ results in a set $F_1$ consisting of at least $k$ points in $B^n$, with all points of $F_1 \cap S$ contained in one of the two open hemispheres into which $H$ divides $S$. This collection of at least $k$ points can be shifted into the interior of $B^n$ by a small translation in the direction perpendicular to $H$. After the shift we have a set $F_2$ in $B^n$, congruent to $F_1$, none of whose points lies on $S$. Then $F_2$ can be expanded within $B^n$ by a homothety centered at $O$, which results in a set $F_3$ of at least $k$ points and more dispersed than $F$. This proves that $\delta_{2k-1}(B^n) < \delta_k(B^n)$. □

The upper bound on $s$ given in the above theorem is probably not the best possible; we conjecture that the best possible is $k-4$, reached in the Rankin equality $\delta_{n+2}(B^n) = \delta_{2n}(B^n)$. Based on the same result, one could also conjecture that, for packing balls in a spherical container in $\mathcal{R}^n$, the relative size of any vacancy in dimension $n$ is bounded above by $\frac{n-2}{n+2}$.

## 3. Packing balls in a cube and in a crosspolytope

We switch from the packing problem to the corresponding dispersion problem: Find the values of $\delta_k(I^n)$. Here again, as it is in the case of the spherical container, the sequence $\{\delta_k(I^n)\}_{k=1}^{\infty}$ is monotonically decreasing, but not strictly. The values of $\delta_k(I^2)$ (for packing $k$ circular disks in a square) are known in a fairly large (but finite) number of cases (see [M], for instance; see also the information and diagrams on the website [P]). Curiously, in none of the known cases a vacancy phenomenon occurs. However, in dimension 3, where only a handful of cases has been solved (see [S,S5,S6,S10]), a vacancy phenomenon occurs relatively soon, namely $\delta_4(I^3) = \delta_3(I^3) = 2\sqrt{2}$ (see Fig. 3, reproduced here from the Internet web site [P] by owner's permission).

**Figure 3.** The vacancy $r_4(I^3) = r_3(I^3)$.

Curiously, no other vacancy phenomenon for ball packings in a cube has been discovered in three dimensions, although it is conjectured that $\delta_{14}(I^3) = \delta_{13}(I^3)$. It is even possible that $\delta_{14}(I^3) = \delta_{12}(I^3)$, as the best known arrangements (see [P]) seem to indicate.

Beginning with dimension 4, an increasingly significant vacancy phenomenon occurs in an infinite sequence of dimensions, related to the existence of Hadamard matrices and to the corresponding vacancy phenomenon for ball packings in the spherical container.

**Definition.** An $n \times n$ matrix $H_n = [a_{i,j}]$ is a *Hadamard matrix* if $|a_{i,j}| = 1$ and $H_n \cdot H_n^T = nI_n$, where $I_n$ denotes the unit $n \times n$ matrix. If an $n \times n$ Hadamard matrix exists, then we say that $n$ is a *Hadamard integer*.

In other words, every entry in $H_n$ is either 1 or $-1$, and the rows of $H_n$ are mutually orthogonal. It is easy to see that if an $n \times n$ Hadamard matrix exists, then $n$ is 1, 2, or a multiple of 4. The well-known, old Hadamard Conjecture asserts that for every $n$ divisible by 4 there is an $n \times n$ Hadamard matrix. Paley's construction (see [BJL], for instance) shows that there exists an $n \times n$ Hadamard matrix when $n$ is divisible by 4 and of the form $2^a(p^m + 1)$, for some positive integers $a$ and $m$, and $p$ an odd prime. Following the recent discovery of Kharaghani and Tayfeh-Rezaie [KTR], the Hadamard Conjecture has been now confirmed for all $n \leq 664$.

Notice that if $n$ is a Hadamard integer, then the set of vertices of the $n$-dimensional cube $I^n$ contains a $2n$-point subset congruent to the set of vertices of a regular $n$-dimensional crosspolytope. Indeed, for any Hadamard matrix $H_n$, the vertices of $I^n$ that are rows of $H_n$ and their antipodes form exactly such a set. (The converse is also true: if the set of vertices of $I^n$ contains the vertices of an $n$-crosspolytope, then $n$ is a Hadamard integer.) This observation leads to the following:

**Theorem 2.** *If $n$ is a Hadamard integer, then $\delta_{2n}(I^n) = \delta_{n+2}(I^n)$.*

*Proof.* Let $B$ be the ball circumscribing $I^n$. Next, let $F$ be a $k$-point subset of $B$, where $n + 2 \leq k \leq 2n$, consisting of vertices of a regular crosspolytope inscribed in $B$. According to the aforementioned result of Rankin, $F$ is maximally dispersed in $B$. Since $n$ is a Hadamard integer, we can assume that $F$ is a subset of $I^n$, and since $I^n \subset B$, the same set $F$ is maximally dispersed in $I^n$ as well. Consequently, $\delta_{2n}(I^n) = \delta_{n+2}(I^n) = \sqrt{2n}$. □

In other words, if $n$ is a Hadamard integer, then in $\mathbb{R}^n$ the same vacancy phenomenon occurs for ball packings in a cube as for ball packings in a sphere. It seems reasonable to expect that in packing balls in a cube, no vacancy exists whose relative size reaches or exceeds 1, just as it is the case for the spherical container.

But the crosspolytope as a container for ball packings allows a vacancy of an unexpectedly high relative size: according to a theorem of Böröczky, Jr. and Wintsche [BW], $\delta_{2n}(X^n) = \delta_4(X^n)$ for every $n \geq 2$. Here we have a vacancy whose relative size tends to infinity linearly with $n$.

In packing congruent cubes, each homothetic to the container, the $n$-cube $I^n$, a vacancy occurs whose relative size tends to infinity with $n$ even faster. Indeed, if two congruent cubes, each homothetic to $I^n$ can be packed in $I^n$, then the small cubes' edge length is at most 1 (an almost trivial fact), and then the container can hold as many as $2^n$ such cubes.

## References


[DH] H. Davenport and G. Hajós, Problem 35 (in Hungarian), *Mat. Lapok* **2** (1951) 68

[AS] J. Aczél and T. Szele, Solutions to Problem 35 (in Hungarian), *Mat. Lapok* **3** (1952) 94–95

[R] R. A. Rankin, The closest packing of spherical caps in $n$ dimensions, *Proc. Glasgow Math. Assoc.* **2** (1955) 139–144

[S] J. Schaer, On the densest packing of spheres in a cube, *Canad. Math. Bull.* **9** (1966) 265–270

[S5] J. Schaer, On the densest packing of five spheres in a cube, *Canad. Math. Bull.* **9** (1966) 271–274

[S6] J. Schaer, On the densest packing of six spheres in a cube, *Canad. Math. Bull.* **9** (1966) 275–280

[S10] J. Schaer, On the densest packing of ten spheres in a cube, *Coll. Math. Soc. János Bolyai* **63** (1991) 403–424

[CS] J. H. Conway and N. J. A. Sloane, Sphere Packings, Lattices and Groups, Springer, 1989

[M] J. B. M. Melissen, Packing and covering with circles, Doctoral Thesis, University of Utrecht, 1997

[BJL] T. Beth, D. Jungnickel and H. Lenz, Design Theory, 2nd ed. rev. Cambridge, England: Cambridge University Press, 1998

[BW] K. Böröczky, Jr. and G. Wintsche, Sphere packing in the regular crosspolytope, *Ann. Univ. Sci. Budapest. Eötvös Sect. Math.* **43** (2001) 151–157

[F99] F. Fodor, The densest packing of 19 congruent circles in a circle, *Geom. Dedicata* **74** (1999) no. 2, 139–145

[F00] F. Fodor, The densest packing of 12 congruent circles in a circle, *Beträge Algebra Geom.* **41** (2000) no. 2, 401–409

[F03] F. Fodor, The densest packing of 13 congruent circles in a circle, *Beträge Algebra Geom.* **41** (2000) no. 2, 401–409

[F04] F. Fodor, The densest packing of 14 congruent circles in a circle, *Stu. Univ. Žilina Math. Ser.* **16** (2003) no. 1, 25–34

[P] H. Pfoertner, Densest Packings of Equal Spheres in a Cube (graphics, summary of conjectures, proven results etc.), web site `http://www.randomwalk.de/sphere/incube/` (2005)

[KTR] H. Kharaghani and B. Tayfeh-Rezaie, A Hadamard matrix of order 428, *J. Combin. Designs* (to appear)

Proc. Int. Conf. – Number Theory and Discrete Geometry No. 6, 2008, pp. 155–163.

# Polynomials Assuming Square Values

M. Ram Murty*

Department of Mathematics, Queen's University, Kingston, Ontario, K7L 3N6, Canada
e-mail: murty@mast.queensu.ca

**Abstract.** If $f(x_1, \ldots, x_n) \in \mathbb{Z}[x_1, \ldots, x_n]$ has the property that every integer specialization gives an integral square value, then $f$ is itself the square of a polynomial. We also give an effective version of this result by using an effective version of a classical theorem of E. Noether along with a theorem of Lang and Weil.

**Keywords.** Polynomials, absolute irreducibility, squares, Lang-Weil theorem, resultants.

**2000 Subject Classification:** Primary 11R09, Secondary 11C08

*In honour of R. P. Bambah on his 80th birthday*

## 1. Introduction

Given a polynomial $f(x_1, \ldots, x_n) \in \mathbb{Z}[x_1, \ldots, x_n]$ with the property that every integer specialization of the $x_i$'s results in a square value, does it follow that $f$ is itself the square of a polynomial? Below, we will show that the answer is yes. In fact, it is not necessary to assume that an infinite number of specializations give rise to a square value. A finite number, depending on the size of the coefficients and the degree of the polynomial suffice and this is our main theorem. The question raises other questions that belong to number theory and algebraic geometry. We will discuss these questions at the end.

The case $n = 1$ of this problem is classical. For example, it appears as a problem in the book by Pólya and Szegö (see p. 132 of [7]). Of course, an analogous result is true for $k$-th powers also. As the referee points out, the multi-variable version of this problem was first investigated by Kojima [4] in 1915. A modern treatment of it can be found in Theorem 52 of [9].

After giving an expository treatment of the single and several variable cases of the problem, we will prove the following effective theorem:

**Theorem 1.** *Let $f(x_1, \ldots, x_n) \in \mathbb{Z}[x_1, \ldots, x_n]$. Then, there is an effectively computable constant $C = C(f)$, depending only on $f$ such that if every integer specialization of $x_1, \ldots, x_n$ with $|x_i| \leq C$ makes $f(x_1, \ldots, x_n)$ a perfect integral square, then $f(x_1, \ldots, x_n)$ is itself the square of a polynomial.*

*Research partially supported by a Natural Sciences and Engineering Research Council (NSERC) grant.

The effectively computable constant $C(f)$ seems to be humongous and depends on the size of the coefficients of $f$, the degree $d$ and $n$. It may be possible by using the work of Deligne, to improve this estimate for $C(f)$, but at present, there are some technical difficulties in this approach. How one may circumvent these difficulties will be addressed in a later paper. However, a refinement of the argument used to prove Theorem 1 will enable us to show:

**Theorem 2.** *Let $f(x_1, \dots, x_n) \in \mathbb{Z}[x_1, \dots, x_n]$ have total degree $d$ and let*

$$k = \binom{n+d-1}{n}.$$

*Let $\|f\|$ denote the sum of the absolute values of all the coefficients of $f$. Put $\psi = 2dt^{2^t}$ where $t = (d+1)(d+2)/2$. If for all integer specializations with $0 \le a_i \le H$ where*

$$H = 2\max(6d\psi, 9(d-1)^2(d-2)^2, 1319007) + k^{2^k} \log 4\|f\|,$$

*we have that $f(a_1, \dots, a_n)$ is a perfect square, then $f(x_1, \dots, x_n)$ is itself the square of a polynomial.*

## 2. Preliminaries

We begin by recalling several results we will need in the later discussion. These are of interest in their own right. The first concerns an old result due to Schur [10]. Given a polynomial $f(x) \in \mathbb{Z}[x]$, we say a prime $p$ is a *prime divisor* of $f$ if $p \mid f(n)$ for some natural number $n$. An excellent introduction to the topic of prime divisors of polynomials can be found in [2].

**Lemma 3 (Schur).** *Let $f(x)$ be a non-constant polynomial with integer coefficients. Then $f$ has infinitely many prime divisors.*

*Proof.* The proof follows Euclid. We induct on the degree of $f$. For polynomials of degree one, this is clear. Let

$$f(x) = a_n x^n + a_{n-1}x^{n-1} + \cdots + a_1 x + a_0.$$

If $a_0 = 0$, we are done by induction, so we may suppose $a_0 \neq 0$. Since $f(x)$ can assume the values $\pm 1$ only a finite number of times, we deduce that $f$ has at least one prime divisor. Suppose there are only finitely many such prime divisors, $p_1, \dots, p_r$ (say). For each natural number $m$, let $N_m = (p_1 \cdots p_r)^m a_0$ and consider

$$f(N_m) = a_0(a_n a_0^{n-1}(p_1 \cdots p_r)^{nm} + \cdots + a_1 a_0 (p_1 \cdots p_r) + 1).$$

For $m$ sufficiently large, the term in the parentheses above is in absolute value greater than 1 and coprime to $p_1 \cdots p_r$. This is a contradiction. □

We also need to recall some basic facts about resultants (see p. 200ff of [5]). Let $R$ be a commutative ring. Given two polynomials

$$f(x) = a_n x^n + a_{n-1}x^{n-1} + \cdots + a_1 x + a_0$$

and

$$g(x) = b_m x^m + b_{m-1} x^{m-1} + \cdots + b_1 x + b_0$$

in $R[x]$, we define the resultant $R(f, g)$ to be the determinant of the $(m+n) \times (m+n)$ matrix

$$\begin{pmatrix} a_n & a_{n-1} & \cdots & a_0 & 0 & \cdots & 0 \\ & a_n & a_{n-1} & \cdots & a_0 & \cdots & 0 \\ & & \cdots & \cdots & & & \\ & & & & a_n & \cdots & a_0 \\ b_m & b_{m-1} & \cdots & b_0 & 0 & \cdots & 0 \\ & b_m & b_{m-1} & \cdots & b_0 & \cdots & 0 \\ & & \cdots & \cdots & & & \\ & & & & b_m & \cdots & b_0 \end{pmatrix}$$

It is not hard to see that there are polynomials $A(x), B(x) \in R[x]$ so that

$$A(x)f(x) + B(x)g(x) = R(f, g).$$

It is well-known that if the coefficients of $f$ and $g$ lie in a field $K$ such that $a_n b_m \neq 0$ and $f, g$ split into factors of degree 1 in $K[x]$, then $R(f, g) = 0$ if and only if $f$ and $g$ have a common root (see p. 203 of [5]). In particular, if $f(x) \in \mathbb{Z}[x]$ has $n$ distinct roots in $\mathbb{C}$, the natural number $R(f, f')$ is non-zero. Thus, if $f(x)$ is squarefree, that is, a product of distinct irreducible polynomials, then $R(f, f')$ is non-zero.

## 3. The case $n = 1$

We can now prove:

**Theorem 4.** *Suppose that $f(x) \in \mathbb{Z}[x]$ is a polynomial such that $f(n)$ is a perfect square for every integer $n$. Then $f(x) = g(x)^2$ for some $g(x) \in \mathbb{Z}[x]$.*

*Proof.* Since $\mathbb{Z}[x]$ is a UFD, we may factor $f(x)$ as a product of irreducible polynomials. By grouping the even powers of the irreducibles occurring in the factorization, we may write $f(x) = g(x)^2 h(x)$ where $h(x)$ is squarefree, that is, a product of distinct irreducible polynomials. Let us suppose that the degree of $h$ is $\geq 1$. By Lemma 3, $h$ has infinitely many prime divisors and so we choose one $p$ which is coprime to $R(h, h')$. Thus, there is a natural number $n$ so that $p | h(n)$. Hence $p | f(n)$. As $f(n)$ is a perfect square, we see that the power of $p$ dividing $f(n)$ is an even power. The same must be true for $h(n)$. Thus, $p^2 | h(n)$. Now, $h(n + tp) \equiv h(n) (\mathrm{mod}\ p)$ so that by the same reasoning, we deduce $p^2 | h(n + tp)$. However, $h(n + tp) = h(n) + pth'(n) (\mathrm{mod}\ p^2)$ so that $p | h'(n)$ if we choose $t$ coprime to $p$ (such as $t = 1$ say). In particular, $p | R(h, h')$ contrary to our choice of $p$. Thus, the degree of $h$ is zero so that $h$ must be a constant. This constant must be a square. □

## 4. The higher dimensional case

Before we begin the discussion of the multi-variable case, it is useful to recall the classical theorem that if $R$ is a UFD then so is $R[x]$. In particular, $\mathbb{Z}[x_1, \ldots, x_n] = \mathbb{Z}[x_1, \ldots, x_{n-1}][x_n]$ is a UFD. Given two polynomials $f, g \in \mathbb{Z}[x_1, \ldots, x_n]$ we may consider them as polynomials in $x_n$ with coefficients in $\mathbb{Z}[x_1, \ldots, x_{n-1}]$. Thus, we may consider the resultant $R_{x_n}(f, g)$ as a polynomial in $x_1, \ldots, x_{n-1}$. By taking an algebraic closure of $\mathbb{Q}(x_1, \ldots, x_{n-1})$, we see that any polynomial factors as a product of linear factors, and we may deduce, as before that $R_{x_n}(f, g) = 0$ if and only if $f$ and $g$ have a common root. In particular, if $f_{x_n} = \partial f/\partial x_n$, then $R(f, f_{x_n}) = 0$ if and only if $f$ and $\partial f/\partial x_n$ have a common root. Thus, if $f$ is squarefree, then the resultant $R(f, f_{x_n}) \neq 0$.

**Theorem 5 (Kojima, 1915).** *Now suppose that we have $f(x_1, \ldots, x_n) \in \mathbb{Z}[x_1, \ldots, x_n]$ having the property that every integer specialization of $x_1, \ldots, x_n$ makes $f(x_1, \ldots, x_n)$ a square. Then, $f(x_1, \ldots, x_n)$ is the square of a polynomial.*

*Proof.* We proceed by induction on the number of variables. As before, we may factor $f = g^2h$, with $h$ a squarefree polynomial. If $h$ has fewer than $n$ variables, we are done by induction. Suppose then that $h$ is a polynomial of $n$ variables. As $h$ is squarefree, the resultant $R(h, h_{x_n})$ (which is a polynomial in $x_1, \ldots, x_{n-1}$) is not identically zero. Thus, we may choose $(x_1, \ldots, x_{n-1}) = (a_1, \ldots, a_{n-1})$ so that the resultant $R := R(h, h_{x_n})(a_1, \ldots, a_{n-1})$ is not zero. Thus, $h(a_1, \ldots, a_{n-1}, x_n)$ is a non-constant polynomial in the single variable $x_n$. For otherwise, its derivative would also be zero and consequently $R$ would be zero which is not the case. By Lemma 3, this polynomial has infinitely many prime divisors. Choose a prime $p$ not dividing $R$. As before, there is an $a_n$ so that $p|h(a_1, \ldots, a_n)$. As $f(a_1, \ldots, a_n)$ is a square, we deduce $p^2|h(a_1, \ldots, a_n)$. The same is true for $h(a_1, \ldots, a_n + p)$. But then,

$$h(a_1, \ldots, a_{n-1}, a_n + p) \equiv h(a_1, \ldots, a_n) + ph_{x_n}(a_1, \ldots, a_n) \pmod{p^2}$$

from which we deduce that

$$p|h_{x_n}(a_1, \ldots, a_n)$$

so that $p|R$, a contradiction. □

## 5. Absolutely irreducible polynomials mod $p$

We now turn to the question of making Theorem 5 effective. The results we invoke form a chapter in classical elimination theory and we refer the reader to pages 177–215 of [11]. We summarise these results below.

Recall that a polynomial $f$ with coefficients in a field $K$ is said to be absolutely irreducible if it is irreducible over the algebraic closure of $K$.

Let $f(x_1, \ldots, x_n) \in \mathbb{Z}[x_1, \ldots, x_n]$ be a polynomial of the form $g^2h$ with $h$ non-constant. Let us note that a polynomial of the form

$$y^2 - f(x_1, \ldots, x_n) \tag{1}$$

with $f(x_1, \ldots, x_n) \in \mathbb{Z}[x_1, \ldots, x_n]$ is irreducible over $\overline{\mathbb{Q}}$. For if it is reducible, it must be a product of two factors which are linear in $y$ and this can only happen only if the discriminant $4f(x_1, \ldots, x_n)$ is a perfect square, which is not the case, by hypothesis. Thus, (1) is absolutely irreducible over $\mathbb{Q}$.

Given a polynomial $f(x_1, \ldots, x_n)$, suppose we can find a prime $p$ such that the number of solutions $(x_1, \ldots, x_n, y) \bmod p$ of the congruence

$$y^2 \equiv f(x_1, \ldots, x_n) (\bmod\ p)$$

is strictly less than $p^n$. Then, there is an $n$-tuple $(a_1, \ldots, a_n)$ so that

$$f(a_1, \ldots, a_n)$$

is not a square mod $p$. Under which conditions can we do this? If the polynomial (1) is absolutely irreducible, that is irreducible over $\overline{\mathbb{F}}_p$, then a famous theorem of Lang and Weil [6] allows us to do this. The question that arises now is if we can find a prime $p$ for which (1) is absolutely irreducible mod $p$.

Given a polynomial $f$ with integer coefficients, we denote by $\|f\|$ the sum of the absolute values of its coefficients. Clearly, $\|fg\| \leq \|f\|\|g\|$. However, a more natural height function for polynomials is given by taking the maximum of the absolute values of the coefficients of $f$ and denoting this by $H(f)$. It is evident that $H(f) \leq \|f\| \leq C(d,n)H(f)$, where $C(d,n)$ is a constant depending on the number of variables and the total degree $d$. In fact, we may take

$$C(d,n) = \binom{d+n}{n},$$

by a simple calculation. Thus, the two heights are comparable.

**Proposition 6 (Gelfond's inequality).**

$$H(f_1) \cdots H(f_r) \leq e^{d_1 + \cdots + d_n} H(f_1 \cdots f_r),$$

*where $d_i$ is the degree in $x_i$ of the product $f_1 \cdots f_r$.*

*Proof.* See page 229 of [3]. □

**Corollary 7.** If $f|g$, then $\|f\| \leq c_1(d,n)\|g\|$ for some effectively computable constant $c_1(d,n)$ depending only on $n$ and the total degree $d$ of $g$.

How is all this relevant to our situation? As before, let us write $f = g^2 h$, with $h$ a squarefree polynomial. Then, by Corollary 7, we have that $\|h\| \leq c_1(d,n)\|f\|$. We will need to use this inequality in applying the following theorem due to E. Noether, in our context.

**Proposition 8 (E. Noether, 1922).** *Let $K$ be a field. Given a polynomial*

$$f(x_1, \ldots, x_n) = \sum_{i_1 + \cdots + i_n \leq d} a_{i_1, \ldots, i_n} x_1^{i_1} \cdots x_n^{i_n} \in K[x_1, \ldots, x_n],$$

*there exist forms $g_1, \dots, g_s$ in variables $A_{i_1,\dots,i_n}$ with $i_1 + \cdots + i_n \leq d$, such that the polynomial $f(x_1, \dots, x_n)$ is reducible over $\overline{K}$ or of degree $< d$ if and only if all the $s$ polynomials $g_j$ vanish when we specialize $A_{i_1,\dots,i_n}$ with $a_{i_1,\dots,i_n}$. Moreover, if*

$$k = \binom{n+d-1}{n},$$

*then the degree of $g_j$ is bounded by $k^{2^k}$. These forms depend only on $n$ and $d$ and are independent of the field $K$ in the sense that if the characteristic of $K$ is zero, they are fixed forms with rational integer coefficients, while if the characteristic of $K$ is a prime $p$, they are obtained by reducing the integral coefficients modulo $p$. In the case characteristic of $K$ is zero, then*

$$\|g_j\| \leq 4^{k^{2^k}}, \quad 1 \leq j \leq s.$$

*Proof.* See page 190 of [11]. □

**Corollary 9.** Let $k$ be as in the previous proposition. Given a polynomial $f(x_1, \dots, x_n) \in \mathbb{Z}[x_1, \dots, x_n]$, which is absolutely irreducible, there exists a natural number $N_f$ such that

$$N_f \leq (4\|f\|)^{k^{2^k}},$$

and $f$ is absolutely irreducible mod $p$ for every prime $p$ coprime to $N_f$.

*Proof.* By Proposition 8, there is a polynomial $g_j$ such that when we specialize to the coefficients of $f$, the value of $g_j$ is a non-zero integer. We let $N_f$ be the smallest of the absolute values of the non-zero values among the $g_j$'s thus obtained. Since the coefficients of $f$ are bounded by $H(f)$, and the degree of $g_j$ is bounded by $k^{2^k}$, we get

$$0 < N_f \leq \|g_j\| \|f\|^{k^{2^k}} \leq (4\|f\|)^{k^{2^k}}.$$

Moreover, for every prime $p$ coprime to $N_f$, we deduce by Noether's theorem that $f$ is absolutely irreducible mod $p$. This completes the proof. □

We will need the following effective version of the theorem of Lang and Weil, proved by Schmidt [11].

**Proposition 10.** *Let $F(x_1, \dots, x_n)$ be a polynomial over $\mathbb{F}_p$ which is of total degree $d > 0$ and absolutely irreducible. Let $N$ be the number of zeros of $F$ in $\mathbb{F}_p^n$. Then,*

$$|N - p^{n-1}| \leq p^{n-2}(\omega(p, d) + 2d\psi),$$

*where*

$$\omega(p, d) = (d-1)(d-2)p^{1/2} + d^2,$$

*and $\psi = 2dt^{2^t}$ with $t = (d+1)(d+2)/2$.*

## 6. Proof of Theorem 1

We apply Corollary 9 to the polynomial

$$F(x_1, \dots, x_n, y) := y^2 - f(x_1, \dots, x_n).$$

Thus, if $p$ is coprime to $N_F$, $F$ is absolutely irreducible mod $p$. By Proposition 10, the number of zeros $N$ mod $p$ satisfies

$$|N - p^n| \le p^{n-1}(\omega(p, d) + 2d\psi).$$

Of these zeros, the number with $y = 0$ can be at most $dp^{n-1}$ since the number of solutions of

$$f(x_1, \dots, x_n) \equiv 0 (\text{mod } p),$$

is bounded by this quantity. So if, we let $N^*$ be the number of zeros of $F$ with $y \neq 0$, we get

$$|N^* - p^n| \le p^{n-1}(\omega(p, d) + 2d\psi + d).$$

In particular, the number of specializations mod $p$ for which $f(x_1, \dots, x_n)$ is a perfect square mod $p$ is

$$\le \frac{1}{2}p^n + \frac{1}{2}p^{n-1}(\omega(p, d) + 2d\psi + d)$$

and this is strictly less than $p^n$ if

$$\omega(p, d) + 2d\psi + d < p.$$

If we choose

$$p > 6d\psi, \quad \text{and} \quad \sqrt{p} > 3(d-1)(d-2),$$

we see that

$$\omega(p, d) + 2d\psi + d < p.$$

So we need $p$ coprime to $N_F$ and $p > \max(6d\psi, 9(d-1)^2(d-2)^2)$. This completes the proof of Theorem 1.

## 7. Proof of Theorem 2

In the proof of Theorem 1, let us consider all specializations of $f(a_1, \dots, a_n)$ with $0 \le a_i \le H$ with $H$ given as in the statement of Theorem 2. We claim that there is a prime $p < H$ which is coprime to $N_F$ and satisfying

$$p > C(d) := \max(6d\psi, 9(d-1)^2(d-2)^2).$$

Indeed, by [12]

$$\theta(H) := \sum_{p<H} \log p > .998684H$$

provided $H \geq 1,319,007$. Since $H = 2(C(d) + \log N_F)$, we see that

$$\sum_{p<H,(p,N_F)=1} \log p > \frac{9}{5}(C(d) + \log N_F) - \log N_F > 0$$

and so there is a prime of the desired type. This completes the proof.

## 8. Concluding remarks

It is clear that the bounds obtained in Theorems 1 and 2 are not optimal. However, Proposition 8 is optimal in the following sense. Ruppert [8] has shown that if $f \in \mathbb{Z}[x, y]$ is absolutely irreducible and has $\deg_x f = m$, $\deg_y f = n$ and height $H(f) = H$, then for any prime $p$ with

$$p > [m(n+1)n^2 + (m+1)(n-1)m^2]^{mn+(n-1)/2} H^{2mn+n-1},$$

the reduction of $f$ mod $p$ is absolutely irreducible. Moreover, if we assume the Bouniakowsky conjecture [1] that predicts that for any irreducible polynomial $g(t) \in \mathbb{Z}[t]$, $g(n)/\delta$ is prime for infinitely many values of $n$ (here, $\delta = \gcd\{g(r) : r \in \mathbb{Z}\}$), then there are infinitely many absolutely irreducible polynomials $f \in \mathbb{Z}[x, y]$ which are reducible mod $p$ where $p$ is a prime with $p > H^{2m}$. Thus, in the case $n = 2$, the power of $H$ in Proposition 8 cannot be substantially improved, modulo the Buniakowsky conjecture, which is generally believed. A closer examination of the proof in [8] shows that what is actually proved is that given $f \in \mathbb{Z}[x, y]$, there is a non-zero natural number $N_f$ satisfying

$$N_f < [m(n+1)n^2 + (m+1)(n-1)m^2]^{mn+(n-1)/2} H^{2mn+n-1},$$

such that $f$ is absolutely irreducible mod $p$ whenever $p$ is coprime to $N_f$. Thus, refining Proposition 8 will not lead to any substantial improvement of our bounds. It may be possible to improve these bounds by other techniques, partly geometric and partly analytic in nature. This will be investigated in a future paper.

Concerning the case of higher powers, the methods extend, in principle. However, getting definitive bounds is not all that straightforward by the methods of this paper. Part of the difficulty is first to establish the absolute irreducibility of the polynomial

$$y^k - f(x_1, \ldots, x_n),$$

when $f(x_1, \ldots, x_n)$ is not a perfect $k$-th power. This is easily done if $k$ is prime and then the argument of this paper easily extends. Thus, one can proceed inductively in this fashion. It is clear that this will again lead to humongous bounds. Consequently, it is thus desirable to think of alternate ways in which better bounds can be obtained.

## Acknowledgements

This paper was first presented at the Chandigarh conference in honour of Professor R. P. Bambah organized by Professors Madhu Raka and R. J. Hans-Gill. I thank them both for their kind hospitality. I would also like to thank Michael Roth, Dinesh Thakur and the referee for their comments on an earlier version of this paper.

## References

[1] V. Buniakowsky, Nouveaux théorèmes relatifs à la distinction des nombres premiers et à la décomposition des entiers en facteurs, *Mém. Acad. Sci. St. Pétersbourg Sci. Math. Phys.* **6** (1857) 305–329

[2] I. Gerst and J. Brillhart, Prime divisors of polynomials, *American Math. Monthly* **78** (1971) 250–266

[3] M. Hindry and J. Silverman, Diophantine geometry, an introduction, *Graduate Texts in Math.* **201** Springer-Verlag, New York, (2000)

[4] T. Kojima, Note on number-theoretical properties of algebraic functions, *Tohoku Mathematical Journal* **8** (1915) Series 1, 24–37

[5] S. Lang, Algebra, Second edition, Addison-Wesley, (1984)

[6] S. Lang and A. Weil, Number of points on varieties in finite fields, *American Journal of Math.* **76** (1954) 819–827

[7] G. Pólya and G. Szegö, Problems and Theorems in Analysis II, translated from the German by C.E. Billigheimer, *Classics in Mathematics* Springer-Verlag, (1998)

[8] W. M. Ruppert, Reducibility of polynomials $f(x, y)$ modulo $p$, *Journal of Number Theory*, **77** (1999) 62–70

[9] A. Schinzel, Polynomials with special regard to reducibility, Cambridge University Press, (2000)

[10] I. Schur, Über die existenz unendlich vielerprimzahlen in einiger speziellen arithmetischen progressionen, *S-B Berlin Math. Ges.* **11** (1912) 40–50

[11] W. Schmidt, Equations over finite fields, An elementary approach, *Springer Lecture Notes* **536** (1976)

[12] J. B. Rosser and L. Schoenfeld, Sharper bounds for the Chebyshev functions $\theta(x)$ and $\psi(x)$, *Math. Comp.* **29** (1975) 243–269

Proc. Int. Conf. – Number Theory and Discrete Geometry No. 6, 2008, pp. 165–169.

# Notes on Prime Number Theorem – III

K. Ramachandra[1,2] and N. K. Sinha[3]

[1]Hon. Vis. Professor, N.I.A.S., IISc Campus, Bangalore 560 012, India
[2]Retired Senior Professor, T.I.F.R Centre, P.O. Box 1234, IISc Campus, Bangalore 560 012, India
[3]Infosys Technologies Limited, Kottara, Mangalore 575 006
e-mail: kram@math.tifrbng.res.in
e-mail: nilotpal_sinha@infosys.com

**Abstract.** We make a case for the study of the real valued function $f(x)$ defined in $x \geq 100$, by

$$x = \int_2^{f(x)} \frac{du}{\log u}.$$

*Dedicated to Professor R. P. Bambah on his eightieth birthday*

## 1. Introduction

We make a case for the study of the real valued function $f(x)$ defined in $x \geq 100$, by

$$x = \int_2^{f(x)} \frac{du}{\log u}. \tag{1}$$

Trivially $f(x)$ is an increasing function of $x$. In a beautiful paper [M.C], M. Cipolla has given an asymptotic expansion of $f(x)$ as $x \to \infty$. The first $m$ terms of the expansion represents $f(x)$ with an error $o(x(\log x)^{-m})$. We quote his result in section §3 of our paper. The integral

$$li(y) = \int_2^{y} \frac{du}{\log u} \ (y \geq 20) \tag{2}$$

is usually called the logarithmic integral.

## 2. Relation of (1) to prime number theory

It is well-known that if $x \geq 100$ and if $\pi(x)$ denotes the number of primes not exceeding $x$, we have

$$\pi(x) = li(x) + O(xExp(-C_1(\log x)^{\frac{3}{5}}(\log\log x)^{-\frac{1}{5}}). \tag{3}$$

This is a very deep result due to the Soviet Mathematician I. M. Vinogradov (see [K.F]). Here $C_1 > 0$ is some absolute constant. On the assumption of Riemann Hypothesis, the error term in (3) is well-known to be $O_\epsilon(x^{\frac{1}{2}+\epsilon})$, for every fixed $\epsilon > 0$. Rieman Hypothesis (RH) is also known to be equivalent to this and also to

$$\sum_{n\leq x}\mu(n) = O_\epsilon(x^{\frac{1}{2}+\epsilon}), \tag{4}$$

for every fixed $\epsilon > 0$. In [K.R.] it was pointed out that if $p_n$ denotes the $n^{th}$ prime number ($p_1 = 2,\ p_2 = 3,\ p_3 = 5,\ldots$) then

$$|p_n - f(n)| \leq \begin{cases} Cn^{\frac{1}{2}}(\log n)^{\frac{5}{2}} \text{ on RH} \\ Cn\ Exp(-D(\log n)^{\frac{3}{5}}(\log\log n)^{-\frac{1}{5}}) \end{cases} \tag{5}$$

the later holding unconditionally. Also it was pointed out that

$$p_n - f(n) = \Omega_\pm((n\log n)^{\frac{1}{2}}\log\log\log n). \tag{6}$$

In (5) $C$ and $D$ are absolute positive constants. In (6) $\Omega_\pm(\ldots)$ represents a quantity $Q$ such that as $n\to\infty$ the upper and lower limits of

$$\frac{Q}{(n\log n)^{\frac{1}{2}}\log\log\log n}$$

are positive and negative respectively. The relation (6) follows from the famous result

$$\pi(x) - li(x) = \Omega_\pm\left(\frac{x^{\frac{1}{2}}}{\log x}\log\ \log\ \log x\right) \tag{7}$$

due to J. E. Littlewood. In [K.R., A.S., K.S.$_1$] it was shown that results like (3) are equivalent to results like

$$\sum_{n\leq x}\mu(n) = O(x\ Exp(-C_2(\log x)^{\frac{3}{5}}(\log\log x)^{-\frac{1}{5}})), \tag{8}$$

where $C_2 > 0$ is an absolute constant.

Also in [K.R., A.S., K.S.$_2$], it was shown that results like

$$\sum_{n_1n_2\ldots n_k\leq x}\mu(n_1)\mu(n_2)\ldots\mu(n_k) = O(x\ Exp(-C_3(\log x)^{\frac{3}{5}}(\log\log x)^{-\frac{1}{5}})), \tag{9}$$

are equivalent to things which resemble (3). Here $C_3 > 0$ depends only on the positive integer parameter $k$ which is a constant.

## 3. Some open questions

In view of the results (5) and (6) it is desirable to obtain an asymptotic expansion for $f(x)$ with an error $O_\delta(x^\delta)$ for every $\delta > 0$ provided we take sufficiently many terms of the expansion. We are also content with the error estimate $O_\epsilon(x^{\frac{1}{2}+\epsilon})$ or even with $O(x\ Exp(-D_1(\log x)^{\frac{3}{5}})(\log\log x)^{-\frac{1}{5}}))$ where $D_1 > 0$ is a constant. Actually $f(x)$ is independent of the prime number functions and it is desirable to study $f(x)$ in its own right. But the present record is the following:

**Theorem (M. Cipolla).** *There exists a sequence $P_j(X)(j \geq 1)$ of polynomials with rational coefficients such that for every non-negative integer m,*

$$f(x) = x\log x + x\log\log x - x + \sum_{j=1}^{m} \frac{(-1)^{j-1} x P_j(\log\log x)}{(\log x)^j} + o(x(\log x)^{-m}). \tag{10}$$

*The polynomials $P_j(x)$ have degree j and leading coefficient $\frac{1}{j}$. In particular*

$$P_1(X) = X - 2, \; P_2(X) = \frac{1}{2}(X^2 - 6X + 11), \ldots. \tag{11}$$

*As a simple corollary, we have,*

$$p_n = n\log n + n\log\log n - n + O\left(\frac{n\log\log n}{\log n}\right) \tag{12}$$

## 4. A remark

We got the reference to Cipolla's paper from [R.J]. Here the author proves

**Theorem ([R.J]).** *We have*

$$(t_{s,n})^{s^{-1}} = p_n + O_r(n(\log n)^{-r}) \; (r \geq 0) \tag{13}$$

*where $t_{s,n}$ is the $n^{th}$ number in the sequence of prime powers $p^t$ (with $t \geq s$) arranged in the increasing order.*

It is not hard to obtain an asymptotic expansion of (2). In fact (by successive integration by parts) we have

$$\int_2^y \frac{du}{\log n} = \sum_{j=1}^{m} \frac{(j-1)!y}{(\log y)^j} + \int_2^y \frac{m!du}{(\log u)^{m+1}} + O(m^m)$$

Here the integral on the right can be broken up into $[2, U]$ and $[U, y]$ and estimated by a suitable choice of $U$(say $U = y^{\frac{1}{2}}$).

## 5. Main result

In the present paper we prove the following theorem.

**Theorem.** *Let $g(u)$, $u \geq 2$ be a positive, continuous increasing function and let $f(x)$ be defined by*

$$x = \int_2^{f(x)} \frac{du}{g(u)}. \tag{14}$$

*Then*

$$f(x) = \int_2^x g(f(u))du + f(2). \tag{15}$$

## 6. Application 1

We restrict to $g(u) = \log u$ so that $f(x) \sim x \log x$. We get

$$f(x) = \int_2^x g(f(u))du + O(1). \tag{16}$$

We write $f(x) = x \log x(1 + o(1))$ so that $\log f(x) = \log(x \log x) + o(1)$. We use this on the RHS of (5) and obtain

$$f(x) = \int_2^x (\log u + \log\log u + o(1))du + O(1). \tag{17}$$

The term $o(1)$ contributes $o(x)$ and the term $\int_2^x (\log u + \log\log u)du$ is easily seen to be

$$x \log x + x \log\log x - x + o(x)$$

Thus

$$f(x) = x \ \log x + x \log\log x - x + o(x) + O(1) \tag{18}$$

using this on the RHS of (5) we get

$$f(x) = \int_2^x \log(u \log u + u \log \ \log u - u + o(u) + O(1))du + O(1). \tag{19}$$

This gives the expansion of $f(x)$ with an error $o\left(\frac{x}{\log x}\right)$. We can continue this process (A times) and proceed as above to get an asymptotic expansion with error $O(x(\log x)^{-A})$ where $A$ is any positive integer constant and obtain the relavent asymptotic expansion. This result is due to M. Cipolla[1]. The authors hope to prove a version of this result which is uniform in $A\left(\leq \frac{2}{3}\frac{\log x}{\log\log x}\right)$.

## 7. Application 2

We can handle $g(u) = (\log u)^r$ where $r$ is any positive constant and obtain the relavent asymptotic expansion.

## 8. Proof of the main result

Differentiating (14) we have

$$1 = \frac{1}{g(f(x))} f'(x) \text{ and so } f'(x) = g(f(x)). \tag{20}$$

Integrating we obtain

$$f(x) - f(2) = \int_2^x g(f(u))du. \tag{21}$$

Transposing we obtain (15).

## Acknowledgements

The authors are very much thankful to Professor Alberto Perelli for sending them a xerox copy of the deep paper by M. Cipolla. The authors are thankful to Professor R. Balasubramanian for drastic simplification of an earlier version of §5.

## References

[M.C] M. Cipolla, La determinazione assintotica dell' $n^{imo}$ numero primo, *Rend. Acad. Sci. Fis. Mat. Napoli* **8(3)** (1902) 132–166

[K.F] Kevin Ford, Vinogradov's integral and bounds for the Riemann zeta-function, *Proc. London. Math. Soc.*, **85(3)** (2002) 565–633

[R.J] Rafael Jakimczuk, Asymptotic formulae, *J. of inequalities in Pure and Appl. Maths.*, vol. 5, Issue 3, Article 62 (2004) 1–8

[K.R.] K. Ramachandra, Notes on Prime Number Theorem – I, Number Theory (ed. Professors R. P. Bambah, (Mrs) R.-J. Hansgill and V. C. Dumir) INSA HBA Publications, (2000) 351–370

[K.R., A.S., K.S.$_1$] K. Ramachandra, A. Sankaranarayanan and K. Srinivas, Ramanujan's Lattice point problem, Prime number theory and other remarks, *Hardy-Ramanujan J.*, **19** (1996) 2–56

[K.R., A.S., K.S.$_2$] K. Ramachandra, A. Sankaranarayanan, and K. Srinivas, Notes on Prime Number Theorm – II, *J. Indian Math. Soc.*, **72**, nos. 1–4, (2005), to appear

Proc. Int. Conf. – Number Theory and Discrete Geometry No. 6, 2008, pp. 171–179.

# Chowla's Problem on the Non-Vanishing of Certain Infinite Series and Related Questions

N. Saradha

School of Mathematics, Tata Institute of Fundamental Research, Homi Bhabha Road, Mumbai 400 005
e-mail: saradha@math.tifr.res.in

*In honour of Professor R. P. Bambah on his 80th birthday*

## 1. Chowla's problem

In 1969, Chowla raised the following question:

*Suppose $f(n)$ is a rational valued, periodic function mod $q$ where $q(>1)$ is a prime number. Then does the infinite series*

$$S = \sum_{n=1}^{\infty} \frac{f(n)}{n} \tag{1.1}$$

*never vanish?*

Throughout the paper, we assume that $f$ is a non-vanishing number theoretic function. Much earlier, in 1949, Chowla [C] himself showed that

$$S \neq 0$$

if $f$ is an odd function and $q$ prime with $\frac{q-1}{2}$ also prime.

Around 1970, Siegel removed the restriction that $\frac{q-1}{2}$ is prime in the result of Chowla. In 1973, Baker, Birch and Wirsing [BBW] solved Chowla's question completely. In fact, Chowla himself solved the question around the same time. The result of Baker, Birch and Wirsing is more general which we state below.

**Theorem A ([BBW]).** *Suppose $f(n)$ is an algebraic valued, periodic function mod $q$. Then the infinite sum $S$ defined in (1.1) does not vanish if $f$ satisfies the following conditions.*

(i) $f(r) = 0$ *if* $1 < \gcd(r, q) < q$
(ii) *The cyclotomic polynomial $\Phi_q$ is irreducible over $\mathbb{Q}(f(1), \ldots, f(q))$.*

The affirmative answer to Chowla's question is immediate since (i) is vacuous when $q$ is prime and (ii) follows when $f$ is rational valued as it is well known that $\Phi_q$ is irreducible over $\mathbb{Q}$. It was shown in [BBW] that conditions (i) and (ii) are necessary. For example, let $q = p^2$ where $p$ is prime and let $f$ be defined by

$$\sum_{n=1}^{\infty} \frac{f(n)}{n^s} = (1 - p^{1-s})^2 \zeta(s)$$

where $\zeta(s)$ is the Riemann Zeta function. For $p = 2$, we get

$$1 - \frac{3}{2} + \frac{1}{3} + \frac{1}{4} + \frac{1}{5} - \frac{3}{6} + \frac{1}{7} + \frac{1}{8} + \cdots = 0$$

with period 4. This shows that (i) is necessary. Let $\chi$, $\chi'$ be quadratic characters mod 12 with conductors 3 and 4, respectively, i.e.,

$$\chi(n) = \begin{cases} 1 & \text{if } n \equiv 1 \pmod 3 \\ -1 & \text{if } n \equiv 2 \pmod 3 \\ 0 & \text{if } n \equiv 0 \pmod 3 \end{cases}$$

and

$$\chi'(n) = \begin{cases} 1 & \text{if } n \equiv 1 \pmod 4 \\ -1 & \text{if } n \equiv 3 \pmod 4 \\ 0 & \text{if } n \equiv 0, 2 \pmod 4. \end{cases}$$

Let $f = 2\chi - \sqrt{3}\chi'$. Then $S = 0$ since

$$L(1, \chi) = \frac{\pi}{2\sqrt{3}}, \quad L(1, \chi') = \frac{\pi}{3}.$$

Here $\mathbb{Q}(f(1), \ldots, f(12)) = \mathbb{Q}(\sqrt{3})$ and $\Phi_{12} = 1 - X^2 + X^4 = (X^2 - \sqrt{3}\,X + 1)$ $(X^2 + \sqrt{3}\,X + 1)$. Thus (ii) is also necessary.

## 2. Transcendental infinite sums

The sum (whenever the series converges) $S$ can be written as a linear form in logarithms of algebraic numbers with algebraic coefficients. More precisely,

$$S = \sum_{n=1}^{\infty} \frac{f(n)}{n} = -\frac{1}{q} \sum_{s=1}^{q-1} \sum_{r=1}^{q} f(r) \xi^{-rs} \log(1 - \xi^s)$$

where $\xi = e^{2\pi i/q}$. See [Le] or [ASST] for a proof. By the famous result of Baker [Ba] on linear forms in logarithms $S$ is either 0 or transcendental. This was observed in [ASST]. From this it follows that

$$L(1, \chi) \text{ is transcendental}$$

for any non-principal character $\chi$ mod $q$. The non-vanishing of any such $L(1, \chi)$ is a basic result in Dirichlet's famous theorem on primes in arithmetic progression. The transcendence of $L(1, \chi)$ for odd characters $\chi$ is also well known for long by the class number formula for quadratic fields. In [ASST] several other infinite series were shown to be transcendental. For instance, series like

$$\sum_{n=1}^{\infty} \frac{1}{(3n+1)(3n+2)(3n+3)}, \quad \sum_{n=1}^{\infty} \frac{F_n}{2^n}$$

represent transcendental numbers. Here $F_n$ denotes the $n$-th Fibonacci number. In the same year when Theorem A appeared, Lehmer investigated series (1) in connection with extended Euler constants. He showed that

$$\sum_{n=1}^{\infty} \frac{1}{(n+1)(2n+1)(4n+1)} = \frac{\pi}{3};$$

$$\sum_{n=1}^{\infty} \frac{1}{(6n+1)\ldots(6n+6)} = \frac{1}{4320}(192 \log 2 - 81 \log 3 - 7\pi\sqrt{3}).$$

Now we know that the above example represents a transcendental number, thanks to the theorem of Baker. Thus the question of non-vanishing of $S$ becomes important. For a survey of these results and other related results, we refer to [ASST], [A], [AS] and [Tij].

## 3. Values connected with the Gamma function

Let us consider the series

$$\sum_{n=1}^{\infty} \frac{f(n)}{n^s}$$

where $f$ satisfies the conditions of Theorem A. This can be written as

$$\sum_{n=1}^{\infty} \frac{f(n)}{n^s} = \sum_{a=1}^{q} f(a) \left\{ \sum_{n=0}^{\infty} \frac{1}{(qn+a)^s} \right\} = \sum_{a=1}^{q} f(a) \frac{1}{q^s} \left\{ \sum_{n=0}^{\infty} \frac{1}{(n+a/q)^s} \right\}.$$

The last series within { } is the well known Hurwitz-Zeta function and we have

$$\sum_{n=1}^{\infty} \frac{1}{(n+a/q)^s} = \frac{1}{s-1} - \frac{\Gamma'(a/q)}{\Gamma(a/q)} + \text{terms involving } (s-1).$$

Thus

$$S = \sum_{n=1}^{\infty} \frac{f(n)}{n} = lt_{s \to 1} \sum_{a=1}^{q} \frac{f(a)}{q} \left\{ \frac{1}{s-1} - \frac{\Gamma'(a/q)}{\Gamma(a/q)} + \text{terms involving } (s-1) \right\}$$

$$= -\frac{1}{q} \sum_{a=1}^{q} f(a) \frac{\Gamma'(a/q)}{\Gamma(a/q)},$$

since $\sum_{a=1}^{q} f(a) = 0$ is a necessary and sufficient condition for the convergence of $S$. By the considerations in Section 2, we get that

$$\sum_{a=1}^{q} f(a) \frac{\Gamma'(a/q)}{\Gamma(a/q)}$$

is transcendental.

The above connection of $S$ with values of $\Gamma'/\Gamma$ was pointed out by Professor M. Ram Murty during the author's lecture in the conference at Chandigarh. This connection has been noticed by Lehmer [Le] in 1973. He has also connected $S$ with some extended Euler constants. This view point was developed in great detail in [RS].

## 4. Necessary and Sufficient condition for the sum $S$ to vanish

In 1982, Okada [O] provided a criterion of all functions $f$ for which (ii) of Theorem A holds and $S = 0$. This criterion is a system of $\varphi(q) + \omega(q)$ homogeneous linear equations in $f(1), \ldots, f(q)$ with rational coefficients where $\varphi(q)$ is the Euler totient function and $\omega(q)$ is the number of distinct prime divisors of $q$. This criterion of Okada was used by Tijdeman [Tij] to show that

(a) $S \neq 0$ if $f$ is completely multiplicative
(b) $S \neq 0$ if $f$ is multiplicative and $| f(p^k) | < p - 1$ for every prime $p$.

In 2003, Saradha and Tijdeman [ST] re-phrased the criterion of Okada so that it is more convenient for application. We give this criterion below. For any integer $n \geq 1$ and any prime $p$, let $v_p(n)$ denote the exact power of $p$ dividing $n$.

**Theorem B ([ST]).** *Let $\Phi_q$ be irreducible over $\mathbb{Q}(f(1), \ldots, f(q))$. Let $M$ be the set of positive integers which are composed of prime factors of $q$ and let*

$$\varepsilon(r, p) = \begin{cases} v_p(q) + \frac{1}{p-1} & \text{if } v_p(r) \geq v_p(q) \\ v_p(r) & \text{otherwise.} \end{cases}$$

*Then $S = 0$ if and only if*

$$\sum_{m \in M} \frac{f(am)}{m} = 0 \text{ for every } a \text{ with } 0 < a < q, \ \gcd(a, q) = 1 \tag{4.1}$$

*and*

$$\sum_{r=1}^{q} f(r)\varepsilon(r, p) = 0 \text{ for every prime divisor } p \text{ of } q \tag{4.2}$$

Applying Theorem B, explicit conditions were given in [ST] under which series like

$$\sum_{n=0}^{\infty} \frac{(-1)^n(an + b)}{(qn + b_1)(qn + b_2)} \text{ or } \sum_{n=0}^{\infty} \frac{an + b}{(qn + s_1)(qn + s_2)(qn + s_3)}$$

do not vanish. Here $a, b$ are algebraic numbers, $b_1, b_2, s_1, s_2$ and $s_3$ are integers.

In his thesis Tengely [Te] has shown that condition (4.1) in Theorem B can be written as a finite sum

$$\sum_{r|q} \varphi\left(\frac{q}{r}\right) f(r) = 0$$

under the additional hypothesis

$$f(m) = f(n) \text{ whenever } v_p(m) = v_p(n) \text{ for every prime } p|q. \tag{4.3}$$

## 5. Erdős problem

The following question is one of the myriad problems posed by Erdős (see [A]):

*Suppose $f$ is a periodic number theoretic function with period $q$ such that*

$$f(n) = \begin{cases} \pm 1 & \text{if } n \not\equiv 0 (\text{mod } q) \\ 0 & \text{otherwise.} \end{cases} \tag{5.1}$$

*Then does the series $S$ vanish?*

The condition $f(q) = 0$ in Erdős problem is necessary as shown by Tengely recently. After a computer search he has shown that $q = 36$ is the least integer for which $S = 0$ when $f(n)$ for $1 \leq n \leq 36$ takes the following sequence of values

$$\{1, -1, -1, -1, -1, 1, 1, 1, -1, 1, -1, -1, 1, -1, 1, -1, -1, 1, 1, 1, -1, 1, -1,$$
$$-1, 1, -1, -1, -1, -1, 1, 1, 1, 1, 1, -1, 1 \dots\}$$

Okada [O] used his criterion mentioned in Section 4 to show that if $f$ satisfies (5.1), then $S \neq 0$ whenever

$$2\varphi(q) \geq q. \tag{5.2}$$

Thus $S \neq 0$ whenever $\omega(q) \leq 2$. In [S], condition (5.2) was relaxed as

$$2\varphi(q) \geq q(1 - \frac{1}{h}) \tag{5.3}$$

where $h = \max_{1 \leq i \leq r}(p_i^{\alpha_i})$ with $q = p_1^{\alpha_1} \dots p_r^{\alpha_r}$. This relaxed condition enabled us to cover more values of $q$ for which $S \neq 0$. For example, it was shown that

$$S \neq 0 \text{ whenever } q \leq 1154 \text{ and } q \neq 525,\ 735,\ 945.$$

The condition (5.3) was used to show that all $f(a)$'s with $\gcd(a, q) = 1$ take the same value. In fact, this holds if (4.3) is assumed.

Thus we show

**Theorem 5.1.** *Suppose $f$ is a periodic number theoretic function with period $q$ such that* (5.1) *holds. Let $q = p_1^{\alpha_1} \cdots p_r^{\alpha_r}$. Assume further that*

$$f(a) = f(b) \text{ whenever } \gcd(a, q) = \gcd(b, q) = 1. \tag{5.4}$$

*Then $S \neq 0$ provided*

$$2\varphi(q) > q\left(1 - \frac{1}{h_0}\right)$$

*where $h_0 = \min_{1 \le i \le r} p_i^{\alpha_i}$.*

*Proof.* Suppose $S = 0$. By Theorem B, (4.2) is valid. By the convergence of $S$, we have

$$\sum_{h=1}^{q} f(h) = 0.$$

Thus

$$\sum_{\substack{h=1 \\ \gcd(h,q)=1}}^{q} f(h) = - \sum_{\substack{h=1 \\ \gcd(h,q)>1}}^{q} f(h). \tag{5.5}$$

By the assumption (5.4)

$$| L.H.S | = \varphi(q). \tag{5.6}$$

Let $i$ be such that $p_i^{\alpha_i} = h_0$. By (4.2), we get

$$\sum_{p_i \| h} f(h) + 2 \sum_{p_i^2 \| h} f(h) + \cdots + (\alpha_i - 1) \sum_{p_i^{\alpha_i - 1} \| h} f(h)$$
$$+ \left(\alpha_i + \frac{1}{p_1 - 1}\right) \sum_{p_i^{\alpha_i} | h} f(h) = 0. \tag{5.7}$$

Further if $p_i \| q$, then $\sum_{p_i | h} f(h) = 0$. From (5.5), (5.6) and (5.7), we get

$$\varphi(q) = \left| - \sum_{\substack{h=1 \\ \gcd(h,q)>1 \\ p_i \nmid h}}^{q} f(h) - \delta_i \left( - \sum_{p_i^2 \| h} f(h) - \cdots - (\alpha_i - 2) \right.\right.$$
$$\left.\left. \times \sum_{p_i^{\alpha_i - 1} \| h} f(h) - \left(\alpha_i - 1 + \frac{1}{p_i - 1}\right) \sum_{p_i^{\alpha_i} | h} f(h) \right) \right|$$

where

$$\delta_i = \begin{cases} 0 & \text{if } p_i \| q \\ 1 & \text{otherwise.} \end{cases}$$

Thus

$$\varphi(q) \le q - \varphi(q) - \frac{q}{p_i} + \delta_i \left( \frac{q}{p_i^2} - \frac{q}{p_i^3} + 2\left( \frac{q}{p_i^3} - \frac{q}{p_i^4} \right) + \cdots \right.$$

$$\left. + (\alpha_i - 2)\left( \frac{q}{p_i^{\alpha_i - 1}} - \frac{q}{p_i^{\alpha_i}} \right) + \left( \alpha_i - 1 + \frac{1}{p_i - 1} \right) \frac{q}{p_i^{\alpha_i}} \right)$$

$$= q - \varphi(q) - \frac{q}{p_i} + \delta_i \left( \frac{q}{p_i^2} + \frac{q}{p_i^3} + \cdots + \frac{q}{p_i^{\alpha_i}} + \frac{1}{p_i - 1} \frac{q}{p_i^{\alpha_i}} \right)$$

$$= q - \varphi(q) - \frac{q}{p_i} + \delta_i \left( \frac{q(p_i^{\alpha_i - 1} - 1)}{p_i^{\alpha_i}(p_i - 1)} + \frac{q}{p_i^{\alpha_i}(p_i - 1)} \right)$$

$$= q - \varphi(q) - \frac{q}{p_i} + \delta_i \left( \frac{q}{p_i(p_i - 1)} \right).$$

Hence

$$2\varphi(q) \le q \left( 1 - \frac{1}{p_i} + \frac{\delta_i}{p_i(p_i - 1)} \right).$$

If $\delta_i = 0$, we have

$$2 \le \frac{q}{\varphi(q)} \left( 1 - \frac{1}{p_i} \right) < 2,$$

a contradiction. Let $\delta_i = 1$. Then $\alpha_i > 1$ and

$$2 \le \frac{2p_i^{\alpha_i}}{p_i - 1} \left( 1 - \frac{1}{p_i} + \frac{1}{p_i(p_i - 1)} \right)$$

implying

$$p_i^{\alpha_i - 1} - 1 \le \frac{p_i^{\alpha_i - 1}}{p_i - 1}$$

which is impossible. □

As a consequence of Theorem 5.1, we get the following corollary.

**Corollary 5.1.** Suppose $f$ satisfies the conditions of Theorem 5.1. Then $S \neq 0$ for all $q < 17325$. Further suppose $\omega(q) = 3$. Then $S \neq 0$ except possibly when $q$ is of the form

$$3^{\alpha_1} 5^{\alpha_2} 7^{\alpha_3} \text{ with } \alpha_1 \ge 3, \alpha_2 \ge 2, \alpha_3 \ge 2 \text{ or}$$

$$3^{\alpha_1} 5^{\alpha_2} 11^{\alpha_3} \text{ with } \alpha_1 \ge 4, \alpha_2 \ge 3, \alpha_3 \ge 2 \text{ or}$$

$$3^{\alpha_1} 5^{\alpha_2} 13^{\alpha_3} \text{ with } \alpha_1 \ge 4, \alpha_2 \ge 3, \alpha_3 \ge 2.$$

*Proof.* We have

$$\frac{\varphi(q)}{q} = \left(1 - \frac{1}{p_1}\right) \cdots \left(1 - \frac{1}{p_r}\right).$$

We check that for $q < 17325$,

$$\frac{\varphi(q)}{q} \geq \frac{1}{2}\left(1 - \frac{1}{h_0}\right). \tag{5.8}$$

Thus by Theorem 5.1, $S \neq 0$ for $q < 17325$. Suppose $\omega(q) = 3$. We may assume by Okada's result, that $\varphi(q)/q \leq 1/2$. Hence we find that

$$(p_1, p_2, p_3) \in \{(3, 5, 7), (3, 5, 11), (3, 5, 13)\}.$$

Let $(p_1, p_2, p_3) = (3, 5, 7)$. Then from (5.8), we get

$$\frac{48}{105} \geq \frac{1}{2}\left(1 - \frac{1}{h_0}\right)$$

which gives $h_0 \leq 11$. Thus $S \neq 0$, whenever

$$q \in \{3 \cdot 5^{\alpha_2} 7^{\alpha_3}, 3^{\alpha_1} 5 \cdot 7^{\alpha_3}, 3^{\alpha_1} 5^{\alpha_2} 7, 3^2 5^{\alpha_2} 7^{\alpha_3}\}.$$

When $(p_1, p_2, p_3) = (3, 5, 11)$ or $(3, 5, 13)$, we get $h_0 \leq 33$ and 65, respectively. This yields the possibilities for $q$ as mentioned in the corollary for which $S$ may vanish.

□

It will be desirable to obtain the improvement of Theorem 5.1 without the assumption (5.4).

## References


[ASST] S. D. Adhikari, N. Saradha, T. N. Shorey and R. Tijdeman, Transcendental infinite sums, *Indag. Math. (N.S.)* **12** (2001) 1–14

[A] S. D. Adhikari, Transcendental Infinite sums and related questions, Number Theory and discrete mathematics, Proceedings of conference, Chandigarh, 2000 (Hindustan Book Agency, 2002), 169–178

[AS] S. D. Adhikari and N. Saradha, Arithmetic nature of sums of certain convergent series, *Resonance*, **7**, no. 11 (2002) 35–46

[Ba] A. Baker, Transcendental Number Theory, Cambridge University Press (1975)

[BBW] A. Baker, B. J. Birch and E. A. Wirsing, On a problem of Chowla, *J. Number Theory* **5** (1973) 224–236

[C] S. Chowla, The Non existence of Non trivial Linear Relations between the Roots of a Certain Irreducible Equation, *J. Number Theory* **2** (1970) 120–123

[Le] D. H. Lehmer, Euler constants for arithmetical progressions, *Acta Arith.*, **27** (1975) 125–142

[O] T. Okada, On certain infinite sums for periodic arithmetical functions, *Acta Arith* **40** (1982) 143–153

[RS] M. Ram Murty and N. Saradha, Transcendental values of the digamma function, *J. Number Theory*, **125** (2007) 298–318.
`http://dx.doi.org/10.1016/j.jnt.2006.09.017`

[ST] N. Saradha and R. Tijdeman, On the transcendence of infinite sums of values of rational functions, *J. London Math. Soc.* **67** (2003) 1–13

[S] N. Saradha, On non vanishing infinite sums, The Riemann Zeta function and Related themes: Papers in honour of Professor K. Ramachandra, Ramanujan Mathematical Society, Lecture Notes Series, **2** (2006) pp. 121–129.

[Te] S. Tengely, Effective Methods for Diophantine Equations, Thesis, Leiden, 2005

[Tij] R. Tijdeman, Some applications of diophantine approximations, Number Theory for the millennium, Proceedings of conference, Urbana, IL, 2000 – Vol III (A. K. Peters, Natick, MA, 2002), 261–284

Proc. Int. Conf. – Number Theory and Discrete Geometry No. 6, 2008, pp. 181–189.

# Covering with Convex Bodies

Gábor Fejes Tóth

Alfréd Rényi Institute of Mathematics, Hungarian Academy of Sciences,
Budapest Pf. 127, 1364 Hungary
e-mail: gfejes@renyi.hu

Almost all problems about packing have their natural dual counterparts for covering. Despite this, compared with the theory of packing, the corresponding theory of covering developed less extensively and only with some delay. While the problem of densest packing of equal balls can be traced to Kepler, and results by Lagrange and Gauss about quadratic forms solve implicitly the problems of densest lattice packing of circles and balls, respectively, the first paper about covering seems to be that of Kershner [37] from the year 1939. The greater interest in packings was due to closer ties to number theory and possible applications in crystalography. However, the disadvantage of the theory of covering has been equalised by the fifties of the last century. The still existing small difference in the state of the two disciplines is due to the fact that problems about covering are generally more difficult than the analogous problems on packing. Here we give a brief survey on the theory of covering whose early development was crucially influenced by the work of Professor R. P. Bambah.

A family $\mathcal{C}$ of convex bodies (compact convex sets with non-empty interior) in $E^d$ is a *covering* if each point of $E^d$ belongs to a member of $\mathcal{C}$. The *density* $d(\mathcal{C})$ of such a covering is defined by a limit that gives the "ratio" of the "total volume of the bodies" to the "volume" of the space covered. The *covering density* $\vartheta(K)$ of a convex body $K$ is the infimum of the densities of all coverings with congruent copies of $K$. If only coverings by translates of $K$ or coverings by lattice arrangements of $K$ are considered, we obtain the analogously defined *translational covering density* $\vartheta_T(K)$ and *lattice covering density* $\vartheta_L(K)$ of $K$, respectively.

In this terminology the above-mentioned theorem of Kershner states that for the unit circle $B^2$ in the plane we have

$$\vartheta(B^2) = \vartheta_L(B^2) = 2\pi/\sqrt{27}.$$

The quantity $2\pi/\sqrt{27}$ is equal to the quotient of the area of a circle to the area of a regular hexagon inscribed into it. A lattice covering realizing the density $2\pi/\sqrt{27}$ consists of the circles circumscribed about the faces of a regular hexagonal tiling.

Around 1950 R. P. Bambah and C. A. Rogers in London and L. Fejes Tóth in Budapest tried to generalise Kershner's result for convex sets instead of circles. Bambah and Rogers considered coverings by translates of a convex disc $K$ and proved a general theorem for finite coverings implying that if $K$ is centrally symmetric then

$$\vartheta_T(K) = \vartheta_L(K). \tag{1}$$

L. Fejes Tóth considered more general arrangements, namely coverings by congruent copies of a convex disc $K$ in which no two discs cross. Two convex discs *cross* if by removing from them their intersection both discs become disconnected. In [29] the following result is announced.

*If non-crossing congruent copies $K_1, \ldots, K_N$ of a convex disc $K$ cover a convex polygon $H$ with at most six sides, then the density of the discs relative to $H$ satisfies the inequality*

$$\sum_{i=1}^{N} \text{area}(K_i)/\text{area}(H) \geq \text{area}(K)/h(K), \tag{2}$$

*where $h(K)$ denotes the maximum area of a hexagon contained in $K$.*

The same bound holds, of course, for the density of a covering of the whole plane. As translates of a convex disc do not cross, it follows that

$$\vartheta_T(K) \geq \text{area}(K)/h(K) \tag{3}$$

for all convex discs $K$. Moreover, the bound in (3) is sharp if $K$ is centrally symmetric. For, according to a theorem of Dowker [16], if $K$ is centrally symmetric, then there is a hexagon of area $h(K)$ inscribed into $K$ and symmetric in the same centre as $K$. This hexagon admits a lattice tiling of the plane, and the corresponding translates of $K$ constitute a lattice covering with density $\text{area}(K)/h(K)$, yielding the relation (1).

One step in the proof of inequality (2) in [29], the construction of generalised Dirichlet cells, contained a gap. The gap was filled by Bambah and Rogers [5].

For centrally symmetric convex discs inequality (1) reduces the determination of $\vartheta_T(K)$ to the determination of $h(K)$. This is a difficult task in general, but for convex $n$-gons Mount and Silverman [41] gave an algorithm determining the value of $h(K)$ in $O(n)$ time.

The assumption that the discs do not cross seems to be superfluous. However, as examples by Heppes and Wegner show, we cannot simply exclude the possibility that crossing pairs occur. Heppes observed that a square is covered by two crossing hexagons, each of them having as vertices the midpoints of the sides of the square and a pair of opposite vertices of the square, but these hexagons cannot cover the square without crossing. Generalising Heppes's construction, Wegner [51] gave, for all $N > 1$, a convex hexagon $H$ and a convex disc $K$ such that $N$ congruent copies of $K$ cover $H$, but they cannot be rearranged so as to obtain a crossing-free covering of $H$.

It was widely believed that such examples exist only for finite arrangements but not for coverings of the whole plane. Recently, W. Kuperberg [40] gave an example of a convex disc $K$ for which he conjectures that its covering density $\vartheta(K)$ cannot be realized by a crossing-free arrangement. To describe his construction, consider a pentagon $P_0 = ABCDE$ such that $AB = AE = \sqrt{3}$, $BC = DE = 1$, $CD = 2$, $\angle BAE = \angle BCD = \angle CDE = 2\pi/3$, and $\angle ABC = \angle DEA = \pi/2$. Three copies

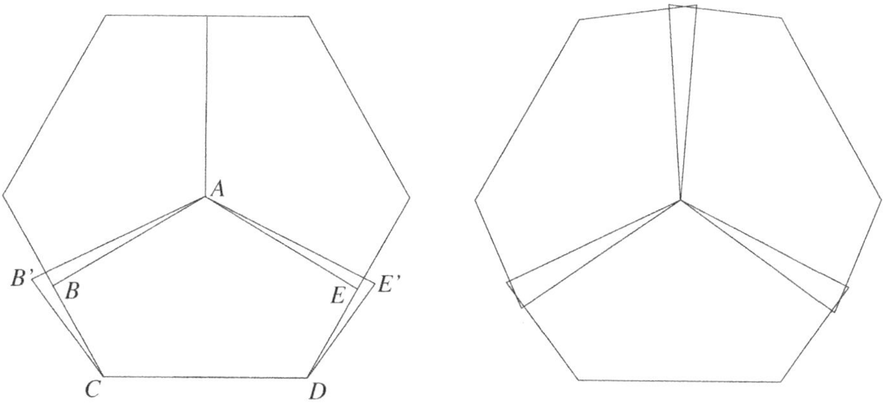


**Figure 1.**

of $P_0$ can be put together to form a regular hexagon $H$ of side-length 2. Displace the vertices $B$ and $E$ into $B'$ and $E'$ so that $AB' = AE' > AB$, $B'C = DE' > BC$ and $\angle AB'C = \angle DE'A < \pi/2$. If $AB' - AB$ and $B'C - BC$ are sufficiently small, then three copies of the resulting new pentagon $P$ can cover $H$ in such a way only that the pentagons cross (Fig. 1). W. Kuperberg conjectures that this occurrence of crossings is inherent to every thinnest covering by congruent copies of $P$.

The use of crossing pairs in a covering is particularly wasteful if $K$ is close to a circle. L. Fejes Tóth suggested that it might be possible to get rid of the crossing-free condition at least for "round" convex discs. A first step in this direction was achieved by Heppes [35]. He called ellipses with semi-axis $a = 1$ and $b \in [0.8561, 1]$ fat, and showed that if a convex hexagon is covered by congruent fat ellipses, then the density of the ellipses relative to the hexagon is at least $2\pi/\sqrt{27}$. The notion of fatness can be extended to convex discs. We say that a convex disc $K$ is *r-fat* if it is contained in a unit circle $C$ and contains a concentric circle $c$ of radius $r$. In [24] it is shown that inequality (3) holds without the non-crossing assumption if $K$ is an 0.933-fat convex disc or an 0.741-fat ellipse.

Before Heppes's result [35] plane-fillers, the circle and obvious modifications of the circle were the only convex discs for which the covering density could be determined. Using the above result on $r$-fat convex discs the covering density of every even-sided regular polygon can be determined.

Another lower bound for $\vartheta_T(K)$ was given by Bambah, Rogers and Zassenhaus [8]. They proved that

$$\vartheta_T(K) \geq \text{area}(K)/2t(K), \tag{4}$$

where $t(K)$ denotes the maximum area of a hexagon contained in $K$. If $K$ is centrally symmetric, then we have $h(K) = 2t(K))$ so that this bound is also exact and it implies (1). However, as it was pointed out in [8], if $K$ lacks a centre of symmetry, then neither (3) nor (4) is best possible, and sometimes one and sometimes the other is stronger.

It is surprising and annoying that the value of $\vartheta_T(K)$ could not be determined for any convex disc $K$ which is not centrally symmetric or a suitable truncation of a centrally symmetric disc. In particular, we do not know the translational covering density of a triangle. Despite this, Bambah and Rogers conjectured that (1) holds not only for the centrally symmetric but for all convex discs $K$. This would follow from the bound

$$\vartheta_T(K) \geq \text{area}(K)/h^*(K),$$

where $h^*(K)$ is the supremum of the areas of centrally symmetric hexagons contained in $K$. The assumption that $K$ is convex is essential in the above conjecture. Bambah, Dumir and Hans-Gill [4] constructed star domains for which (1) does not hold.

A common generalisation of the bounds (3) and (4), which brings a bit closer to the above conjecture of Bambah and Rogers, is given in [22]. It uses the notion of $p$-hexagon introduced in [38] by W. Kuperberg. A *p-hexagon* is a convex hexagon with a pair of parallel opposite sides of equal length. Opposite sides are which are separated by exactly two other sides in each direction. Pentagons with a pair of parallel sides, quadrilaterals and triangles are considered to be degenerate $p$-hexagons. Let $\hat{h}(K)$ be the maximum of the areas of $p$-hexagons contained in $K$. Then we have

$$\vartheta_T(K) \geq \text{area}(K)/\hat{h}(K). \tag{5}$$

It is possible that (2) holds without the assumption that the discs do not cross, and that even the inequality

$$\vartheta(K) \geq \text{area}(K)/h(K)$$

holds. This would imply the conjecture of L. Fejes Tóth that

$$\vartheta(K) = \vartheta_L(K)$$

if $K$ is centrally symmetric. In connection with this conjecture we should note that the property that $\vartheta(K) = \vartheta_L(K)$ is rather exceptional among convex sets. As it was shown in [27], if $\mathcal{C}$ denotes the space of all convex discs in the plane equipped with the Hausdorff metric and $\mathcal{N}$ is the subset of $\mathcal{C}$ consisting of all convex discs $K$ for which $\vartheta(K) \neq \vartheta_L(K)$, then $\mathcal{N}$ is an open dense set in $\mathcal{C}$.

It is likely that the coverings of the plane by congruent copies of a convex discs $K$ realizing the minimal density $\vartheta(K)$ show some regularity. In particular, it can be conjectured that the density $\vartheta(K)$ can be achieved by a periodic arrangement. This is not the case if $K$ is not convex. A construction of Schmitt [49] yields a star-shaped set $K$ for which $\vartheta(K)$ cannot be realized by a periodic covering. (in [49] packings are considered, but the same construction works for coverings as well).

There are various results concerning coverings of bounded sets by a finite number of translates or non-crossing congruent copies of a convex discs $K$ which implies inequality (3) or (4). Bambah and Rogers obtained inequality (1) as a special case of a rather complicated theorem giving a lower bound for the density of a finite number of translates of a centrally symmetric convex discs relative to the polygon determined by the centres of the "outer discs". A proof of their theorem appeared in [8]. In [11] it is shown that (2) remains valid if instead of a polygon with at most six sides a convex disc is covered, provided $N$ is sufficiently large. In [21] it is proved that if $H$ is an arbitrary convex disc and $K_1, \ldots, K_N$ are translates of a centrally symmetric convex disc $K$ covering $H$, then (2) holds whenever $N \geq 26$. It is conjectured that here the condition on $N$ can be weakened to $N \geq 2$. This is confirmed in [28] if $K$ is a circle. In this latter case, even the condition that $H$ is convex can be relaxed: inequality (2) holds if $K$ is a unit circle, $n \geq 2$ and $H$ is the complement of the union of circles of radius 16.7437 (see [23] Lemma 4).

Inequality (2) remains true under more general conditions. Already L. Fejes Tóth noted that instead of congruent convex discs we can take affinely equivalent convex disc with the same area. Böröczky observed that the condition that the discs have equal area can be relaxed. Let $F_K(n)$, $n = 3, 4, \ldots$ denote the maximum area of a $n$-gon contained in $K$. Inequality (2) holds if $K_1, \ldots, K_N$ are affine images of a convex disc $K$ such that

$$\frac{\text{area}(K_i)}{\text{area}(K_j)} \geq \frac{F_K(7) - F_K(6)}{F_K(6) - F_K(5)}.$$

Unfortunately, Böröczky did not publish his proof. A rather complicated proof for the special case for covering by circles is given in [10]. A proof of Böröczky's result appeared first as a special case of a more general result in [20].

All the results mentioned so far were generalisations of Kershner's result concerning covering with equal circles. We now mentions a generalisation of this result in another direction. Let $a$ and $p$ be positive numbers satisfying the isoperimetric inequality $p^2/a \geq 4\pi$. Let $\mathcal{D}(a, p)$ be the class of all discs with area not less than $a$ and perimeter not greater than $p$. What is the thinnest covering of the plane by discs from $\mathcal{D}(a, p)$? A *regular arc-sided $n$-gon* is a convex disc obtained from a regular $n$-gon $P$ by joining each consecutive vertices of $P$ by congruent circular arcs. It is conjectured that a thinnest covering of the plane by discs from $\mathcal{D}(a, p)$ arises from a regular hexagonal tiling of appropriate edge-length by joining equal circular segments to the tiles, so as to obtain regular arc-sided hexagons of area $a$ and perimeter $p$. In [25] it is shown that this is indeed the case if the convex hulls of the discs do not cross. If $p^2/a = 4\pi$, then, of course, the class $\mathcal{D}(a, p)$ consists of a single circle, and we get Kershner's result.

For a triangle $T$ we have $\vartheta_L(T) = 3/2$. According to a theorem of Fáry [19] we have

$$\vartheta_L(K) \leq \frac{3}{2}$$

for all convex discs $K$, that is, the convex disc which admits the least economic lattice covering is the triangle. W. Kuperberg [38] showed that every convex disc $K$ contains a $p$-hexagon of area $(2\sqrt{3} + 3)/8$. As $p$-hexagons tile the plane, it follows that

$$\vartheta(K) \leq \frac{8(2\sqrt{3} - 3)}{3} = 1.2376043\ldots$$

for all convex discs $K$. Elaborating on Kuperberg's method Ismailescu [36] showed that

$$\vartheta(K) \leq 1.2281771\ldots,$$

which is the presently best known upper bound for the covering density of general convex discs. A result of Sas [48] implies that

$$\vartheta(K) \le \frac{2\pi}{\sqrt{27}} = 1.2091995\ldots$$

for every centrally symmetric convex disc $K$. Equality can hold only for an ellipse. It is conjectured that equality holds for all ellipses, but we know this only for the 0.741-fat ellipses.

We mentioned in the introduction that the densest lattice packing of balls (in three dimensions) was determined by Gauss. The dual counterpart of this result was established by Bambah [1]. We have

$$\vartheta_L(B^3) = \frac{5\sqrt{5}\pi}{24}.$$

Alternative proofs were given by Barnes [9] and Few [32]. A proof of the conjecture that $\vartheta(B^3) = 5\sqrt{5}\pi/24 = 1.4635\ldots$ seems to be out of reach, but the bound $\vartheta(B^3) \ge 3\sqrt{3}(\arccos(1/3) - \pi)/2 = 1.4322\ldots$ by Coxeter, Few and Rogers [12] comes quite close to the conjectured value. Bambah and Woods [7] proved that the density of a double lattice covering by balls in $E^3$ cannot be smaller than $5\sqrt{5}\pi/24$. A double lattice arrangement arises as the union of a lattice arrangement and a translate of it.

In [2] Bambah gave an example of a lattice covering by balls in four dimensions of density $2\sqrt{5}\pi^2/25$ which was confirmed to be the thinnest one by Delone and Ryškov [15]. The lattice covering density $\vartheta_L(B^5) = 245\sqrt{35}\pi^2/3888$ of the ball in five dimensions was determined by Ryškov and Baranovskiĭ [47].

No lower bound for $\vartheta(K)$ is known for a class of general convex bodies in dimensions higher than 2. The first lower bounds for $\vartheta(B^d)$ by Erdős and Rogers [17] and by Bambah and Danenport [3] were of the form

$$\vartheta(B^d) \ge c - \varepsilon_d$$

with $c = 16/15$ and $c = 4/3$, respectively and $\varepsilon_d \to 0$ as $n \to 0$. An essential improvement on these bounds was achieved by Coxeter, Few and Rogers [12]. They showed that

$$\vartheta(B^d) \ge \tau_d \sim \frac{d}{e\sqrt{e}} \qquad (d \to \infty)$$

which is the best known lower bound for $\vartheta(B^d)$ even today. The geometric meaning of $\tau_d$ is the following: it is the ratio between the total volume of the sectors of $d+1$ unit balls centred at the vertices of a regular simplex of edge-length $\frac{1}{2}\sqrt{2(d+1)/d}$ and the volume of the simplex.

Early results concerning the existence of economic lattice coverings of $E^d$ by convex bodies by Rogers [42,44,45], Bambah and Roth [6], Davenport [13,14] and Watson [50] established bounds of the form

$$\vartheta_L(K) \le c^d,$$

the smallest value of $c = 1.017$ given by Watson. In [46] Rogers proved that

$$\vartheta_L(K) \leq d^{\log_2 \ln d + c}$$

for some constant $c$ and every convex body $K$ in $E^d$. For the $d$-dimensional ball he obtained the stronger bound

$$\vartheta_L(B^d) \leq cd(\ln d)^{\frac{1}{2}\log_2 2\pi e}$$

for some constant $c$. Gritzmann [34] extended the last bound for every convex body $K$ which has an affine image symmetric in at least $\log_2 \ln d + 4$ coordinate hyper-planes.

If we do not restrict to lattice arrangements, much more economic coverings can be found. Rogers [43] showed that

$$\vartheta_T(B^d) \leq d \ln d + d \ln \ln d + 5d$$

for all convex bodies in $E^d$. This compares quite favourably with the lower bound on $\vartheta(B^d)$ by Coxeter, Few and Rogers.

We close by mentioning a result [26] showing that for many convex bodies $\vartheta(K)$ cannot be realized by lattice coverings. If $d \geq 3$, then for every strictly convex body $K$ there exists an affine image $K'$ of $K$ such that $\vartheta(K') < \vartheta_L(K')$. Moreover, if $\mathcal{C}$ is a periodic covering of $E^d$ with translates of a strictly convex body $K$, then there exists a covering $\mathcal{C}'$ with congruent copies of a suitable affine image $K'$ of $K$ such that $\mathcal{C}'$ is thinner than $\mathcal{C}$.

## References


[1] R. P. Bambah, On lattice coverings by spheres, *Proc. Nat. Inst. Sci. India* **20** (1954) 25–52

[2] R. P. Bambah, Lattice coverings with four-dimensional spheres, *Proc. Cambridge Philos. Soc.* **50** (1954) 203–208

[3] R. P. Bambah and H. Davenport, The covering of $n$-dimensional space by spheres, *J. London Math. Soc.* **27** (1952) 224–229

[4] R. P. Bambah, V. C. Dumir and R. J. Hans-Gill, Covering by star domains, *Indian J. Pure Appl. Math.* **8** (1977) 344–350

[5] R. P. Bambah and C. A. Rogers, Covering the plane with convex sets, *J. London Math. Soc.* **27** (1952) 224–229

[6] R. P. Bambah and K. F. Roth, A note on lattice coverings, *J. Indian Math. Soc. (N.S.)* **16** (1952) 7–12

[7] R. P. Bambah and A. C. Woods, The thinnest double lattice covering of three-spheres, *Acta Arith.* **18** (1971) 321–336

[8] R. P. Bambah, C. A. Rogers and H. Zassenhaus, On coverings with convex domains, *Acta Arith.* **9** (1964) 191–207

[9] E. S. Barnes, The covering of space by spheres, *Canad. J. Math.* **8** (1956) 293–304

[10] G. Blind, Über Unterdeckungen der Ebene durch Kreise, *J. Reine Angew. Math.* **236** (1969) 145–173

[11] K. Jr. Böröczky, Finite packing and covering by congruent convex domains, *Discrete Comput. Geom.* **30** (2003) 185–193

[12] H. S. M. Coxeter, L. Few and C. A. Rogers, Covering space with equal spheres, *Mathematika* **6** (1959) 147–157

[13] H. Davenport, Sur un systeme de spheres qui recouvrent l'espace $n$ dimensions, *C. R. Acad. Sci. Paris* **233** (1951) 571–573

[14] H. Davenport, The covering of space by spheres, *Rend. Circ. Mat. Palermo* (2) **1** (1952) 92–107

[15] B. N. Delone and S. S. Ryškov, Solution of the problem of the least dense lattice covering of a 4-dimensional space by equal spheres (Russian), *Dokl. Akad. Nauk SSSR* **152** (1963) 523–524 = *Soviet Math. Dokl.* **4** (1963) 1333–1334

[16] C. H. Dowker, On minimum circumscribed polygons, *Bull. Amer. Math. Soc.* **50** (1944) 120–122

[17] P. Erdős and C. A. Rogers, The covering of $n$-dimensional space by spheres, *J. London Math. Soc.* **28** (1953) 287–293

[18] P. Erdős, and C. A. Rogers, Covering space with convex bodies, *Acta Arith.* **7** (1961/1962) 281–285.

[19] I. Fáry, Sur la densité des réseaux de domaines convexes, *Bull. Soc. Math. France* **78** (1950) 152–161

[20] G. Fejes Tóth, Covering the plane by convex discs, *Acta Math. Acad. Sci. Hungar.* **23** (1972) 263–270

[21] G. Fejes Tóth, Finite covering by translates of centrally symmetric convex domains, *Discrete. Comput. Geom.* **2** (1987) 353–363

[22] G. Fejes Tóth, Note to a paper of Bambah, Rogers and Zassenhaus, *Acta Arith.* **50** (1988) 119–122

[23] G. Fejes Tóth, Covering the plane with two kinds of circles, *Discrete Comput. Geom.* **13** (1995) 445–457

[24] G. Fejes Tóth, Covering with fat convex discs, *Discrete. Comput. Geom.* **34** (2005) 129–141

[25] G. Fejes Tóth and A. Florian, Covering of the plane by discs, *Geom. Dedicata* **16** (1984) 315–333

[26] G. Fejes Tóth and W. Kuperberg, Thin non-lattice covering with an affine image of a strictly convex body, *Mathematika* **42** (1995) 239–240

[27] G. Fejes Tóth and T. Zamfirescu, For most convex discs thinnest covering is not lattice-like, in: *Intuitive Geometry*, ed. by K. Böröczky and G. Fejes Tóth, Colloquia Math. Soc. János Bolyai **63**, North-Holland, Amsterdam – New York (1994) 105–108

[28] L. Fejes Tóth, Über dichteste Kreislagerung und dünnste Kreisüberdeckung, *Comment. Math. Helvetici* **23** (1949) 342–349

[29] L. Fejes Tóth, Some packing and covering theorems, *Acta Univ. Szeged, Acta Sci. Math.* **12/A** (1950) 62–67

[30] L. Fejes Tóth, *Regular Figures*, Pergamon Press, Oxford, (1964)

[31] L. Fejes Tóth, *Lagerungen in der Ebene auf der Kugel und im Raum*, Zweite verbesserte und erweiterte Auflage, Springer-Verlag, Berlin–New York, (1972)

[32] L. Few, Covering space by spheres. *Mathematika* **3** (1956) 136–139

[33] Florian, A. Packing and covering with convex discs, *Studia Sci. Math. Hungar.* **21** (1986) 251–265

[34] P. Gritzmann, Lattice covering of space with symmetric convex bodies, *Mathematika* **32** (1985) 311–315

[35] A. Heppes, Covering the plane with fat ellipses without the non-crossing assumption, *Discrete Comput. Geom.* **29** (2003) 477–481

[36] D. Ismailescu, Covering the plane with copies of a convex disc, *Discrete Comput. Geom.* **20** (1998) 251–263

[37] R. Kershner, The number of circles covering a set, *Amer. J. Math.* **61** (1939) 665–671

[38] W. Kuperberg, Packing convex bodies in the plane with density greater than 3/4, *Geom. Dedicata* **13** (1882) 149–155

[39] W. Kuperberg, Covering the plane with congruent copies of a convex body, *Bull. London Math. Soc.* **21** (1989) 82–86

[40] W. Kuperberg, Personal communication

[41] D. M. Mount and R. Silverman, Packing and covering the plane with translates of a convex polygon, *J. Algorithms* **11** (1990) 564–580

[42] C. A. Rogers, A note on coverings and packings. *J. London Math. Soc.* **25** (1950) 327–331

[43] C. A. Rogers, A note on coverings, *Mathematika* **4** (1957) 1–6

[44] C. A. Rogers, Lattice coverings of space with convex bodies. *J. London Math. Soc.* **33** (1958) 208–212

[45] C. A. Rogers, Lattice covering of space: The Minkowski-Hlawka theorem. *Proc. London Math. Soc.* (3) **8** (1958) 447–465

[46] C. A. Rogers, Lattice coverings of space, *Mathematika* **6** (1959) 33–39

[47] S. S. Ryškov and E. P. Baranovskiĭ Solution of the problem of the least dense lattice covering of five-dimensional space by equal spheres (Russian), *Dokl. Akad. Nauk. SSSR* **222** (1975) 39–42 = *Soviet Math. Dokl.* **16** (1975) 586–590

[48] E. Sas, On a certain extremum-property of the ellipse, (Hungarian) *Mat. Fiz. Lapok* **48** (1941) 533–542

[49] P. Schmitt, Discs with special properties of densest packings, *Discrete Comput. Geom.* **6** (1991) 181–190

[50] G. L. Watson, The covering of space by spheres. *Rend. Circ. Mat. Palermo* (2) **5** (1956) 93–100

[51] G. Wegner, Zu einem ebenen Über deckungs problem, *Studia Sci. Math. Hungar.* **15** (1980) 287–297

Made in the USA
Middletown, DE
05 November 2026

36603416R00119